Kaironomia

Leonardo, *The Deluge*, c. 1514. (Copyright reserved. Reproduced by gracious permission of Her Majesty The Queen.)

Kaironomia

On the Will-to-Invent

ERIC CHARLES WHITE

Cornell University Press

ITHACA AND LONDON

CORNELL UNIVERSITY PRESS GRATEFULLY ACKNOWLEDGES
A GRANT FROM THE ANDREW W. MELLON FOUNDATION
THAT AIDED IN BRINGING THIS BOOK TO PUBLICATION.

First published 1987 by Cornell University Press.

International Standard Book Number 0-8014-1993-X
Library of Congress Catalog Card Number 86-24235
Printed in the United States of America
Librarians: Library of Congress cataloging information appears on the last page of the book.

The paper in this book is acid-free and meets the guidelines for permanence and durability of the Committee on Production Guidelines for Book Longevity of the Council on Library Resources.

Contents

Prologue

The art of rhetoric has traditionally held that eloquence in the expression of thought requires the effective use of tropes and figures, of argumentative strategies and forms of discourse. This insistence on the role technique plays in the process of persuasion does not imply, however, that eloquent speech can be achieved simply by manipulating a fixed repertoire of conventional devices. Prior to the structural arrangement and ornamentation of an argument is its initial conception. And invention, or that part of rhetorical theory concerned with the genesis of ideas, comprises not merely the selection of persuasive means from an existing body of prototypes but the creation of these means as well. Indeed, rhetoric's traditional inventory of topics for argument constitutes a history of past instances of original contrivance. Eloquence is inspired by the need to produce a novel response to the situation confronting the speaker, by the desire to give voice to the previously "unheard of"—precisely, by the "will-to-invent," which must therefore be recognized as a crucial point of reference orienting any philosophic view of rhetoric. What, then, is the will-to-invent? Why is it so important to be inventive? And how might the mind be assured an abundance of new ideas? These are questions this essay will attempt to answer. *Kaironomia* has a polemical motive: to the extent that it would actively promote a stance toward experience facilitating invention, it can be understood as an encomium, or formal praise, of the will-to-invent.

As an encomium celebrating the will-to-invent, *Kaironomia* will employ the three varieties of rhetorical proof—*ethos, logos,* and *pathos*—commonly recognized as appropriate in this form of literary discourse. For its logical proof, *Kaironomia* will draw upon a diverse array of philosophic traditions to establish the validity of its thesis that invention must constantly be renewed. Although respective traditions of thought will be put to roughly equivalent argumentative use, each will remain irreducibly distinct. *Kaironomia* will offer multiple perspectives on the will-to-invent, perspectives that complement without excluding or subsuming one another. In this way, it will avoid ratifying some one magisterial discourse whose insistence on having the last word would foreclose the possibilities of invention. Intending an effect of plural orchestration rather than theoretical resolution or closure, *Kaironomia* would prove by its own example that inventiveness depends on flexibility with respect to the interpretive vocabularies in terms of which the mind construes experience.

At the same time, however, the dispersion of logical significance among heterogeneous philosophic traditions will be contained within the bounds of a purely imaginative unity when *Kaironomia* has recourse to its two other modes of proof, *ethos* and *pathos*. Briefly, where *ethos* designates a persuasive strategy based on personal charisma, *pathos* involves appealing to audience predispositions in order to secure commitment to the speaker's point of view. *Kaironomia* will thus portray the *ethos* of the will-to-invent as a vortex of creative energy for which, as pathetic proof, a venerable lineage will be discovered in received values and assumptions. Taken together, *Kaironomia*'s use of *ethos* and *pathos* would render the prospect of endless interpretation so delightfully compelling that the reader will henceforth adopt the will-to-invent as an ideal orienting the course of mental evolution. Demonstrative argument concerning the objective need for continually renewed invention will be suffused, then, with subjective yearning for an ideally improvisational activity of making sense. *Kaironomia* will transpire as *eroticized thought*. It will practice a middle style, midway between disinterested investigation and poetic rapture, between what is the case and what ought to be, midway, finally, between reality and desire.

This intermingling of logical, aesthetic, and moral persuasion requires that *Kaironomia* be received in a certain spirit. Claiming for itself no conclusive preeminence and hoping, in fact, to serve as provocation to further invention, the present essay is meant to be read and enjoyed as dismissable speculation.

ERIC CHARLES WHITE

Ithaca, New York

PART ONE

The Paradox of the Liar

Every time we enter a crisis the absurdity is total; understand that dialectics can only set our closet in order in moments of calm. You know very well that at the high point of a crisis we always work by impulse, just the opposite of foresight, doing the most unexpected and wildest sort of thing. And at that moment precisely it could be said that there was a sort of saturation of reality, don't you think? Reality comes on fast, it shows itself with all its strength, and precisely at that moment the only way of facing it is to renounce dialectics. . . . Reason is only good to mummify reality in moments of calm or analyze its future storms, never to resolve a crisis of the moment. But these crises are like metaphysical outbursts, like a state that perhaps if we hadn't chosen the path of reason, would be the natural and current state of *Pithecanthropus erectus.*

—JULIO CORTAZAR, *Hopscotch*

Kairos is an ancient Greek word that means "the right moment" or "the opportune." The two meanings of the word apparently come from two different sources. In archery, it refers to an opening or "opportunity" or, more precisely, a long tunnel-like aperture through which the archer's arrow has to pass. Successful passage of a *kairos* requires, therefore, that the archer's arrow be fired not only accurately but with enough power for it to penetrate. The second meaning of *kairos* traces to the art of weaving. There it is the "critical time" when the weaver must draw the yarn through a gap that momentarily opens in the warp of the cloth being woven.[1] Putting the two meanings together, one might understand *kairos* to refer to a passing instant when an opening appears which must be driven through with force if success is to be achieved.

Since the circumstances enabling success may change at any time, *kairos* implies that there can never be more than a contingent and provisional management of the present opportunity. Success depends, in other words, on adaptation to an always mutating situation. Understood as a principle of invention, or a prescription concerning the way thought should encounter reality, one might say that *kairos* therefore counsels thought to act always, as it were, on the spur of the moment. Such an activity of invention would renew itself and be transformed from moment to moment as it evolves and adapts itself to newly emergent contexts. The fluid and relative moments of the immediate situation would be constitutively involved in the invention process, which would become an improvisational one, a more or less successful attempt to "make do" with whatever is conveniently at hand. *Kairos* thus establishes the living present as point of departure or inspiration for a purely circumstantial activity of invention.

1. R. B. Onians, *The Origins of European Thought* (Cambridge: Cambridge University Press, 1951), pp. 343–49.

This improvisational readiness to chance the fortune of the moment entails a conception of temporality according to which the flow of time is understood as a succession of discontinuous occasions rather than as duration or historical continuity. That is, instead of viewing the present occasion as continuous with a causally related sequence of events, *kairos* regards the present as unprecedented, as a moment of decision, a moment of crisis, and considers it impossible, therefore, to intervene successfully in the course of events merely on the basis of past experience. How can one make sense of a world that is eternally new simply by repeating the ready-made categories of tradition? Tradition must answer to the present, must be adapted to new circumstances that may modify or even disrupt received knowledge. Rather than understand the present solely in terms of the past, one should seek to remain open to an encounter with the unforeseen spontaneity we commonly describe as the "mother of invention." *Kairos* discovers in every new occasion a unique opportunity to confer meaning on the world. It is, in that sense, an appropriate term with which to initiate an essay on the will-to-invent.

Gorgias of Leontini, a sophist of the Greek Enlightenment with whose philosophy of rhetoric this introductory section of *Kaironomia* will largely be concerned, privileged *kairos* as the master concept in his rhetorical theory. The concept of *kairos* has in fact been said to underlie not only Gorgias's rhetorical theory but his ethics and aesthetics as well.[2] For Gorgias, *kairos* stands for a radical principle of occasionality which implies a conception of the production of meaning in language as a process of continuous adjustment to and creation of the present occasion, or a process of

2. Mario Untersteiner, *The Sophists,* trans. K. Freeman (Oxford: B. Blackwell, 1954), p. 160; the other principal sources for this discussion of Gorgias are W. K. C. Guthrie, *The Sophists* (Cambridge: Cambridge University Press, 1971); George Kennedy, *The Art of Persuasion in Greece* (Princeton: Princeton University Press, 1963); Jacqueline de Romilly, *Magic and Rhetoric in Ancient Greece* (Cambridge, Mass.: Harvard University Press, 1975). See also: Marcel Detienne and Jean-Pierre Vernant, *Cunning Intelligence in Greek Culture and Society,* trans. Janet Lloyd (Atlantic Highlands, N.J.: Humanities Press, 1978). Detienne and Vernant's fascinating study of "adaptable cunning" as a major preoccupation in Greek religious thought suggests a traditional basis for Gorgias's philosophy of rhetoric.

continuous *interpretation* in which the speaker seeks to inflect the given "text" to his or her own ends at the same time that the speaker's text is "interpreted" in turn by the context surrounding it. This subordination of meaning to occasion entails the view that the truth of an utterance is immanent and not anterior to the situation of the utterance itself. In other words, the persuasive force of a speech does not derive from its correspondence to a preexistent reality or truth. Truth is relative to the speaker and the immediate context. There is no criterion or standard that might once and for all found and orient the progress of knowledge. The persuasive force of the truth must be renewed at each occasion and cannot become, therefore, a routine accomplishment. Falling prey to the vagaries of historical circumstance, knowledge comes to depend on the occasion of utterance. Or to put it another way, there is no meaning outside of a specific context of rhetorical persuasion.

An epistemological dilemma first broached in Greek tragic drama may provide a philosophical basis for the importance Gorgias attaches to the occasion of thought. As has often been remarked, in the structure of feeling characteristic of tragedy the mind finds itself suspended in anguished irresolution between two mutually exclusive alternatives. Tragedy induces in its audience a state of cognitive dissonance, an unresolved tension between opposite or contrary forces. The condition of mind that tragic consciousness presupposes does not respect the logic of identity, of the excluded middle.[3] Tragic consciousness sets before itself two irreconcilable ways of viewing human being without finally endorsing either one or the other outlook. The philosophical consequence that the sophists of the Greek Enlightenment extrapolated from this tragic irresolution took the form of the doctrine, traditionally attributed to Protagoras, that "in every experience there are two *logoi* in opposition to each other."[4] Inaugurating a diacritical logic founded on a

3. Jean-Pierre Vernant, "Greek Tragedy: Problems of Interpretation," in Richard Macksey and Eugenio Donato, eds., *The Structuralist Controversy* (Baltimore: Johns Hopkins University Press, 1970), pp. 273–95; for an extended discussion of the cultural setting in which tragic drama emerged, see A. W. Adkins, *Merit and Responsibility* (Oxford: Clarendon Press, 1960).

4. Untersteiner, quoting Diogenes Laertius, p. 19.

tension between contraries, the doctrine of the *dissoi logoi* proposes that truth is self-contradictory, or riven by the strife of opposites. The tragedy of knowledge resides in the fact that two contradictory statements can be made on any issue.

The enervating effect of this unending strife can be overcome, according to Gorgias, only in the pragmatic closure thought achieves when it applies itself to a particular context. *Kairos* represents, in this sense, Gorgias's attempt to make the best of the unresolvable dilemma that ensues from the lack of coincidence between Being and knowledge, the inability of knowledge to give definitive expression to the essence of Being. A decision made on the spur of the moment, on the basis, that is, of *kairos,* would resolve the epistemological dilemma by pure force of will, trusting to the fortune of the moment to produce an utterance that truly answers to its occasion.[5]

Such a decision, however, could never be more than a provisional one, because the tragic recognition that speech can express only a partial truth means that on some other occasion a possibility presently denied will have in turn its moment of truth. While on the one hand, the doctrine of the *dissoi logoi* stands for a theory of knowledge, on the other hand, then, it proposes a view of reality itself according to which the historical unfolding of reality can be expected to assume the form of an unending flux. The doctrine of the *dissoi logoi* implies an ontology not unlike the one put forward by Heraclitus, who saw reality as an "ever-living Fire" that "rests by changing," or remains the same by becoming other than itself.[6] As the familiar phrase has it, "everything flows and nothing abides, everything gives way and nothing stays fixed." The ever-changing flux of appearances harbors no permanent or univocal structure of meaning beneath its surface. Reality is only this ever-variegating surface, and lacks the founding principle that would orient its muta-

5. In *The Lives of the Sophists,* Philostratus reports: "Gorgias [seems] to have begun extemporaneous oratory. For coming into the theatre of the Athenians he had the boldness to say 'suggest a subject', and he was the first to proclaim himself willing to take this chance, showing apparently that he knew everything and would trust to the moment to speak on any subject"; see Rosamond Kent Sprague, ed., *The Older Sophists* (Columbia: University of South Carolina Press, 1972), p. 31; a translation of Diels-Kranz, eds. *Die Fragmente der Vorsokratiker* (1934).

6. Untersteiner makes this connection between the doctrine of the *dissoi logoi* and Heraclitean process reality; p. 24.

tions toward some predetermined end. The thought that seeks to take advantage of the present opportunity must resign itself, therefore, to its own historically contingent character. As a prescription concerning the basis on which thought can *begin* to intervene in the world, *kairos* is offered with the understanding, then, that thought must always be willing, as circumstances change, to begin *again.*

An aspiration for a mode of thought which would always be in the process of beginning again figures prominently in the writings of Gertrude Stein. There is, in fact, a striking resemblance between Stein's desire for an activity of consciousness in which the mind would never finally rest on an object but stay always, as it were, in motion, and Gorgias's promotion of *kairos* as the principle determining rhetorical invention. Since Gorgias never prepared a systematic treatise on the art of the right moment, a brief excursion into Stein's conception of invention may therefore help to enlarge the implications of *kairos.*

Recommending a practice of speculative thought alert to its own occasion, which she refers to in this context as "at the same time talking and listening," Stein writes:

> The trouble with a great many so-called intelligent people is they mix up remembering with talking and listening, and as a result they have theories about anything but as remembering is repetition and confusion, and being existing that is listening and talking is action and not repetition intelligent people although they talk as if they knew something are really confusing, because they are so to speak keeping two times going at once, the repetition time of remembering and the actual time of talking but, and as they are rarely talking and listening, that is the talking being listening and the listening being talking, although they are clearly saying something they are not clearly creating something, because they always are remembering, they are not at the same time talking and listening.[7]

In other words, the impulse to repeat, which ratifies the comfortable notion that knowledge is in principle finite, encourages us to speak not *to* the present occasion (the locus of genuine novelty) but *from* (in

7. Gertrude Stein, *Lectures in America* (New York: Random House, 1935), p. 180.

imitation of) the purely ideal constructs of our memories. Instead of a mode of utterance which recurs to its own practice, taking account of the circumstances from which it issues, this "repetition compulsion" would have us ignore the context of utterance, the "suchness" that would otherwise force the recognition that the world is eternally new, eternally different. In order to break this grip of memory on thought, Stein advocates a discourse responsive to its context, responsive to the vagaries of historical process. Were we to "speak in pieces and say no matter," as she elsewhere puts it, we would disrupt the routine accumulation of meaning according to a logic of linear continuity that forecloses the possibility of creativity. That is, we would no longer stand witness before the unhappy spectacle, to borrow a phrase from James Joyce, of "the rite words by the rote order." Stein would thus move her audience from living according to the rituals of memory to the alertness and mobility of "being existing." Consciousness would then no longer rotate around memory, but would be open instead to the irrational novelty of the moment.

The recurrence in the writings of Gertrude Stein of an invention theory based on *kairos* can be understood to represent Stein's radicalization of the view her teacher William James took of the "stream of consciousness." Following Heraclitus, for whom the chief characteristic of reality resides in its mutability, James held that the inchoate flux of the world is nonsensical prior to the imposition of a human fabric of meaning: "Out of what is in itself an indistinguishable, swarming *continuum,* devoid of distinction or emphasis, our senses make for us, by attending to this notion and ignoring that, a world full of contrasts, of sharp accents, of abrupt changes, of picturesque light and shade."[8] The world of "picturesque light and shade" which we ordinarily *know* is, according to James, a purely conventional artifice reductive of the immense variety of experience: "The mind is at every stage a theatre of simultaneous

8. William James, *The Principles of Psychology* (1890; rpt. New York: Dover, 1950), 1: 284. Stein shares James's view that reality is essentially nonsensical; in *Tender Buttons,* she writes: "real is, real is only, only excreate a no since," a phrase interpreted by Richard Bridgman to mean that "reality is the creative excretion of the moment, which may be regarded as non-sense"; see *Gertrude Stein in Pieces* (New York: Oxford University Press, 1970), p. 131.

possibilities. Consciousness consists in the comparison of these with each other, the selection of some, and the suppression of the rest by the reinforcing and inhibiting agency of attention." Since the way we attend to experience is largely a matter of habit (we are, James says, "mere walking bundles of habits"), the shape of the world we perceive has already been determined in advance of our perceiving it: "We find it quite impossible to disperse our attention impartially over a number of impressions. . . . We actually *ignore* most of the things before us." And yet, contrary to this pessimistic view that sees us habituated to a fixed and therefore misleading perception of an essentially mutable reality, James can say: "*My experience is what I agree to attend to.*" Dogmatic repetition is not inevitable: "When the identical fact recurs, we *must* think of it in a fresh manner, see it under a somewhat different angle, apprehend it in different relations from those in which it last appeared."[9] Although the stream of consciousness can eddy back upon itself in moments of self-reflection, it can never seize hold of itself in its entirety. Always other in relation to its previous states, thought encounters the world from ever-different vantages.

On the basis of this more hopeful side of James's thinking, Stein will attempt to break out of the fixed priorities of memory or habit in order to unleash a processual dynamic of thought subversive of every point of arrest. She will seek to loosen her attachment to the idea of "reality" so that she can initiate a practice of attending to the present ensuing not in a given once-and-for-all representation of the world (the subject/object universe traditionally imposed in order to stabilize the swarming continuum of experience) but in an endless production of sense, or nonsense, since such meaning, surpassing received knowledge, would have to emerge as paradox. Stein explains her intention in strikingly contemporary terms when she defines her task as a *decentering* of attention. For Stein, the notion of "the center" stands for those habits of attention, or remembering (as she says, "custom is in the center"), that lead to repetition and confusion instead of at the same time talking and listening. She advocates a nomadic or wandering "center" that is not constrained

9. James, 1: 288, 127, 284, 402, 233.

within the limits of a habitual circuit.[10] Were we to truly "act so that there is no use in a center," we would disperse our attention across the entire range of our impressions. Such a mobile center would be responsive to the fluctuating pulse of the present occasion. Stein's "central" tenet, that "any occasion shows the best way," thus intends a stance toward experience which would no longer operate in the shadow of memory.[11]

The sophist who, like Gorgias, would take the present occasion as point of departure for rhetorical invention must be similarly suspicious of memory, or the conventional wisdom the Greeks called "opinion." With *kairos,* Gorgias intends to disclose what received knowledge would sooner forget: *kairos* entails a recognition of the antithetical nature of the truth, and consequently, of the provisional character of the *logos* of the moment. In this sense, Gorgias would have been inconsistent ever to have codified his views on *kairos;* to do so would have meant falling prey to his own memory or opinion. A bare initiatory principle, a principle of spontaneity and risk, *kairos* can never form the basis of a systematic treatise on the management of the opportune. Since *kairos* stands for precisely the irrational novelty of the moment that escapes formalization, any science of "kaironomy" would find itself incoherently promising foreknowledge of chance.[12] To put it another way, no treatise on the occasional nature of utterance could be itself exempt from occasionality, or the inevitability of its own supersession. If every occasion presents a unique challenge to the situational, context-oriented consciousness of the sophist, then the sophist's interpretive ingenuity will nowhere find itself resumed in a definitive statement. Gorgias's

10. See Bridgman's discussion of *Tender Buttons* in *Gertrude Stein in Pieces,* pp. 124–36; Bridgman remarks there that "rather than existing at a fixed point, the center moves and is everywhere and nowhere" (p. 132).

11. All of the quotations in this paragraph are from *Tender Buttons;* see *The Selected Writings of Gertrude Stein,* ed. Carl Van Vechten (New York: Random House, 1962), pp. 459–509.

12. "Kaironomy" would thus approximate to "Pataphysics," Alfred Jarry's term for the mock-serious science that would examine the laws governing exceptions; which brings to mind another standpoint from which *kairos* can be understood: Tristan Tzara held that "everything in Dada stresses its *occasional,* circumstantial nature, the real aim of art being integration with the present-day world"; see Motherwell, ed., *The Dada Painters and Poets* (1951; rpt. New York: G. K. Hall, 1981), p. 397.

philosophy of rhetoric will coincide with the trajectory of his career, an amoebic expansion whose track at any given moment foresees no final destination. The rhetorical practice of the sophist who allows *kairos* to figure in the invention of speech will issue, then, in an endlessly proliferating style deployed according to no overarching principle or rational design. The orator who invents on the basis of *kairos* must in fact always go beyond the bounds of the "rational" to the extent that this "will-to-spontaneity" succeeds in evading the burden of the past, the repertoire of collective norms that dissemble the ambiguous, ambivalent nature of reality.

Gorgianic rhetoric finds itself opposed on this point by the Peripatetic view that the orator's argument should be secured by taking public opinion as the source for its major premise. This view is not, however, an exclusively Aristotelian one. Within the sophistic movement it was advocated, for instance, by Protagoras, whose "man the measure" phrase had the effect of making the orator finally dependent on conventional wisdom.[13] A reliance on the *doxa* was a leading characteristic not only of Aristotelian but of Isocratean rhetoric and, therefore, of the mainline of rhetorical speculation in antiquity. This invocation of collective values and assumptions indicates confidence in their ultimate validity on the part of theorists less dubious of conventional wisdom than Gorgias. But recourse to the *doxa* must also be recognized as a purely expedient strategy for many practicing sophists of the day. With reference to the motives that may have impelled Aristotle to take up rhetorical theory, Paul Ricoeur suggests that rhetoric was originally a weapon for triumphing over one's opponents, a technique for winning agreement by any means available. The orator's rhetorical "bag of tricks" enabled him to exercise power by playing upon the fears and prejudices of his auditors. As a body of assumptions firmly held even though untested against the claims of experience, the *doxa* is especially vulnerable to cynical manipulation. The public's unreflecting adherence to received knowledge renders them easily victimized by an unscrupulous speaker.

Concerned about the problem of deception in rhetoric, Aristotle

13. In this connection, Guthrie notes that Protagoras's skepticism stops short because "the appearance of the moment is subordinated to a higher standard, the end or purpose of human nature and society" (p. 187).

sought to reform contemporary rhetorical practice. His *Rhetoric* is meant to cleanse the art of its characteristic abuses by distinguishing legitimate persuasion from flattery and intimidation, "the subtlest forms of violence."[14] Aristotle reforms rhetoric, Ricoeur says, by placing it under the supervision of philosophy. He thus theorizes a mode of logical proof—the enthymeme or rhetorical syllogism—which provides at least a probable representation of reality. And in his treatment of appropriate styles of discourse, the various kinds of ethical and pathetic proof, as well as in his enumeration of *topoi* or generally held beliefs—all the typical resources of contemporary rhetorical practice—Aristotle provides for their philosophically sanctioned employment. A responsible orator schooled in Aristotelian rhetoric would be able to state his case persuasively before an audience untrained in the rigors of philosophic argument without engaging in self-serving trickery. In this sense, Aristotle's *Rhetoric* follows the course mapped out by Plato in the *Phaedrus* when Socrates argues in effect that rhetoric is properly a handmaiden to morality and politics. Gorgias and Aristotle are equally suspicious, then, of the deceptive character of persuasion based on conventional wisdom, differing only in that where Aristotle is willing to accede to the *doxa* so long as it is philosophically supervised, Gorgias would paradoxically confound the predispositions of his audience. But this relatively minor difference regarding the means by which each proposes to overcome the dogmatism of everyday life actually portends a far more profound disagreement.

Despite his willingness to theorize a rhetorical practice in which argumentative proof is less than certain, Aristotle did not share Gorgias's skepticism concerning the possibility of final truth. Indeed, Aristotle's elaboration of a probability rhetoric in which persuasion depends on appealing to public opinion can be regarded as a corollary of his commitment to the Platonic aspiration for certain knowledge. Book 1 of the *Rhetoric,* in which the enthymeme is distinguished from the syllogism on the ground that it depends on a merely probable premise, presupposes Book 2, in which Aristotle

14. Paul Ricoeur, *The Rule of Metaphor: Multi-disciplinary Studies in the Creation of Meaning in Language,* trans. Robert Czerny with Kathleen McLaughlin and John Costello (Toronto: University of Toronto Press, 1977), p. 11.

discusses the psychology and typical beliefs of representative members of the general audience. As an orator, Aristotle may speak the language of opinion, but his appraisal of popular culture has a philosophic basis. He is thus assured of his persuasiveness: Aristotle *knows* which commonplaces to invoke, what manner of ethical and pathetic proof will secure his audience's sympathetic attention. The realm of the merely probable in Aristotle's *Rhetoric* is circumscribed within a more extensive realm of certain knowledge: knowledge of the audience, of human nature.

The Aristotelian orator's access to the truth influences not only his tactical calculations with respect to particular groups in society but the fundamental import of his speech as well. Aristotle's emphasis on the enthymeme as a mode of proof encourages the philosophically oriented rhetorician to recognize as his principal obligation the following task: notwithstanding his audience's inability to comprehend the intricacies of strictly logical deduction, he must *tell the truth*. Aristotle would finally dispel the possibility of deception by bringing rhetoric's discourse of probability into close alignment with philosophy's discourse of certainty. His confidence that his own logical system could convey the essence of Being has, then, important consequences for the *Rhetoric*. As many commentators have noted, the *Rhetoric* is an ambiguous work. Even as he is pragmatically willing to employ nonlogical modes of proof, Aristotle remains hostile to them. On the one hand, an eminently practical and high-minded manual of persuasive technique seeking to eliminate self-serving trickery and deceit, on the other, the *Rhetoric* is frequently impatient with the *doxa* and eager to dispense with those forms of persuasion not based on Aristotelian logic. As Jacqueline de Romilly observes, Aristotle's reform of contemporary rhetorical practice entails a "promotion toward reason and austerity." Following Aristotle's description of rhetoric as the counterpart of dialectic, "most of the traditional parts of rhetoric were gradually absorbed by logic and thereby lost their importance."[15] Aristotle's rhetorical theory contrasts strikingly in this respect with that of Gorgias. If Aristotle would correct a naive, indeed, deluded world view with

15. de Romilly, pp. 66, 70.

reference to a philosophically sophisticated definition of reality, Gorgias would avoid positing any putatively conclusive meaning and would be willing to affirm a provisional truth only so long as it was accompanied by an awareness of the ultimately one-sided character of every philosophic outlook.

The views of Gorgias and Aristotle are thus at the furthest possible remove from one another. Their respective positions in fact define between them the space in which an important controversy of the Greek Enlightenment was played out, a controversy concerning at once the possibility of knowledge and the nature of reality. To put the matter briefly: where Gorgias holds that there is no absolutely knowable reality, Aristotle maintains that the substance of a thing can indeed be determined; and where Gorgias practices a "plural" style, or one emphasizing the figural resources of language, Aristotle generally will advise against extensive use of the flowers of rhetoric. This contrast, an antithetical one at the level of both form (the question of an appropriate style) and content (the question of the nature of reality), could not be made by comparing Aristotle with Protagoras, for the radical nature of Protagoras's skepticism is tempered by his willingness to presuppose "Man" as the arbiter of values. That is, although Protagoras did not credit the existence of ultimate truth, he was willing, unlike Gorgias, to depend on "common sense." Alternatively, the contrast could not be made by comparing Gorgias with Plato, for Plato, unlike Aristotle, does not hesitate to employ a "plural" style. In the *Phaedrus,* for instance, Plato uses figurative language in Socrates' poetic fable on the ascent of the soul because he is dubious of the claim that philosophic discourse has some privileged purchase on the truth, indeed that any mode of utterance can truly deliver the essence of the thing to which it refers. Plato relies on the ambiguities and the pathetic force of figurative language to gesture toward the transcendent without pretending to definitively represent what must in the end remain unrepresentable. Furthermore, although there is in Plato a desire for the apprehension of a timeless visionary truth, Plato nevertheless remains committed to the endless interpretive interaction of dialogue. In the second half of this introductory section of *Kaironomia,* then, the specifically philosophic dimension of Aristotle's rhetorical theory will provide the backdrop for a discussion of Gorgias's rhetorical practice in his treatise "On Not-Being" and the "Encomium on Helen," a backdrop that should enable some further implications of Gorgias's notion of *kairos* to emerge in sharp relief.

The epistemological basis of Aristotle's *Rhetoric* can be traced to his revision of the Platonic theory of forms, Plato's hypothesis, following Parmenides' doctrine of Being, of a permanent and univocal structure of meaning behind the flux of appearances. When Aristotle gave up Plato's theory of suprasensible forms, he did so precisely in order to ensure the possibility both of a stable truth immanent in nature and of a language capable of expressing it. Aristotle's first step, logically at least, was to transfer the forms from the transcendental realm to the world of sense experience, with the result that the forms became "indwelling" in experience.[16] In his *Metaphysics,* Aristotle holds that this solution marks a culmination in metaphysical speculation: "The question that was asked long ago, is asked, and is always a matter of difficulty, 'What is being?' is the question 'What is substance?'" (1082b 2). And for Aristotle, there need be no discrepancy between the "substance" of a thing and the philosophic discourse that would communicate that substance because, as he puts it in *On Interpretation,* "Just as all men have not the same writing, so all men have not the same speech sounds, but the mental experiences, which these sounds directly symbolize, are the same for all, as also are those things of which our experiences are the images" (16a 4–8). In a remarkable foreshadowing of representational theories of language, Aristotle maintained, according to Porphyry in his commentary on Aristotle's Categories: "As things are, so are the expressions which primarily indicate them."[17]

16. This characterization of Aristotle's revision of the Platonic theory of forms is borrowed from John Jones, *On Aristotle and Greek Tragedy* (London: Chatto and Windus, 1962); Guthrie says that although Aristotle "gave up the transcendence of the Platonic forms, he continued to believe in the existence of permanent substances or essences corresponding to universal terms," and in that sense still affirmed Plato's hypothesis of "a reality beyond and independent of our knowledge and beliefs" (pp. 53, 184).

17. G. B. Kerferd, "Aristotle," in Paul Edwards, ed., *The Encyclopedia of Philosophy* (New York: Macmillan, 1967), I:156. All of the Aristotle quotations are from *The Works of Aristotle,* ed. W. D. Ross (Oxford: Clarendon Press, 1908–62).

Aristotle's theory of knowledge can therefore be said to imply a two-step invention process consisting, first, of the discovery of a structure already, in fact, eternally there in the world, and subsequently, of its expression in language. In the *Posterior Analytics,* Aristotle provides a description in effect of his conception of invention:

> So out of sense perception comes to be what we call memory, and out of frequently repeated memories of the same thing develops experience; for a number of memories constitute a single experience. From experience again—i.e. from the universal now stabilized in its entirety within the soul, the one beside the many which is a single identity within them all—originates the skill of the craftsman and the knowledge of the man of science, skill in the sphere of coming to be and science in the sphere of being. (100a 5–9)

For Aristotle, Plato's forms, always immanent in nature, reveal themselves as such when they are abstracted from the sensible world and embodied in human representations. Philosophic discourse would therefore comprise a body of statements that are logically mimetic of reality (just as the structure of tragic drama, according to Aristotle in the *Poetics,* is experientially mimetic of the human situation). For a philosophic statement truly to achieve its purpose of communicating the substance of a thing, the units of meaning in language must exist in a stable and univocal relationship with their objects. This aspiration for exact correspondence between language and reality, crucial if language is really to tell the truth, will determine the strictures on style that Aristotle lays down in Book 3 of the *Rhetoric,* especially his preference for a relatively unadorned plain style and attendent characterization of the figural resources of language as primarily ornamental.

Aristotle commences his discussion of style with a reminder that style is of secondary importance in relation to the subject matter of a speech: "The first question to receive attention was naturally the one that comes first naturally—how can persuasion be produced from the facts themselves. The second is how to set these facts out in language" (*Rhetoric* 1403b 17–18). Since persuasion from the facts themselves is of primary importance, Aristotle values that style most that interferes the least with an audience's access to the facts. The chief criterion of stylistic excellence for Aristotle is, therefore, lucidity. As Jacqueline de Romilly remarks, the "style that [Aristotle] recommends should be transparent."[18] And since Aristotle values a logically clear style over every other, he counsels against the frequent use of tropes and figures drawn from poetry. Aristotle regards the poetic style as an archaic remnant from a time when the respective functions of poetry and oratory were not clearly distinguished—

> Now it was because poets seemed to win fame through their fine language when their thoughts were simple enough, that the language of oratorical prose at first took a poetical color, e.g., that of Gorgias. Even now most uneducated people think that poetical language makes the finest discourses. That is not true: the language of prose is distinct from that of poetry. (*Rhetoric* 1404a 24–27)

In contrast to the language of poetry, with its figures and tropes that underscore the disparity between word and object, the language of prose directly delivers the logical architecture of reality. Although Aristotle does provide for the use of figurative language, the context in which it is appropriate (impassioned speech, as in a peroration) is narrowly circumscribed. Figurative language may be employed to enforce the persuasiveness of the argument of a speech only so long as it does not hinder a clear understanding of the facts of the matter. For Aristotle, pleasing ornament must not obscure plain sense, the train of enthymematic reasoning. His stylistic ideal of clarity without meanness and distinctiveness without extravagance attaches greatest importance, then, to the transmission of logical content.

Paul Ricoeur has argued in a subtle analysis of Book 3 of the *Rhetoric* that Aristotle's characterization of tropes and figures as rhetorical ornament issues from the belief that his own logical system constitutes a fully adequate instrument for representing reality.

18. de Romilly, p. 73. See also G. M. A. Grube's discussion of Aristotle's theory of style in *The Greek and Roman Critics* (London: Methuen, 1965).

If Aristotelian logic does indeed convey the essence of Being, then every other way of depicting the world can serve at most a decorative purpose. Aristotle implies the ornamental nature of figurative language in this definition of metaphor from the *Poetics:* "Metaphor consists in giving the thing a name that belongs to something else; the transference being either from genus to species, or from species to genus, or from species to species, or on grounds of analogy" (1457b 7–9). Since it transpires within an already constituted unity (precisely, the order of things for which Aristotelian logic offers a privileged description), metaphor cannot be regarded as a figure productive of knowledge. In the typical case, Ricoeur says, Aristotle's theory of metaphor posits a deviation from ordinary usage in which a term is borrowed from one domain and substituted for another term on the basis of a perception of likeness or hidden resemblance between semantic fields. Ricoeur observes that "if the metaphorical term is really a substituted term, it carries no new information, since the absent term (if one exists) can be brought back in; and if there is no new information conveyed, then metaphor has only an ornamental, decorative value."[19] As a purely decorative effect, metaphor can contribute no new meaning to the orator's discourse. The relationship that it discovers has already been predicted by Aristotelian logic. The representational possibilities of metaphor and figurative language in general are thus subordinate to the perspective on the world afforded by Aristotle's logical system, which subsumes every other point of view. Since the order of things, according to the theory of substance, is eternally given just as it is, only one style of discourse—Aristotle's own—can truly correspond to it. All others deviate from this realistic norm.

But what if, as Gorgias believed, the world is constantly changing, in a condition of ceaseless flux that extends to the tragic strife of irreconcilable contraries within knowledge itself? How could the finite resources of Aristotle's logical system, which presuppose a corresponding predictable universe, equip an orator to grapple successfully with infinitely various occasions? In this connection, Ricoeur argues that another view of figurative language is implicit

19. Ricoeur, p. 20.

but undeveloped in Aristotelian thought, a view that attributes an inventive potential to tropes and figures. Ricoeur remarks that Aristotle's definition of metaphor as the substitution of terms from diverse semantic fields involves the notion of "categorical transgression, understood as a deviation in relation to a preexisting logical order, as a dis-ordering in a scheme of classification." According to Ricoeur, when Aristotle describes metaphor as transgressive, he unwittingly renews the possibilities of invention: "Should we not say that metaphor destroys an order only to invent a new one; and that the category-mistake is nothing but the complement of a logic of discovery?" This line of reflection, Ricoeur notes, is "completely opposed to any reduction of metaphor to a mere 'ornament.' Pushing this thought to the limit, one must say that metaphor bears information because it 'redescribes' reality." Ricoeur goes on to attribute an inventive potential to tropes and figures: "If metaphor belongs to an heuristic of thought, could we not imagine that the process that disturbs and displaces a certain logical order, a certain conceptual hierarchy, a certain classification scheme, is the same as that from which all classification proceeds?"[20] Aristotelian logic would thus be the product of one metaphorically constituted semantic field, one among many possible ways of describing reality, all of which have their origin in the figurative resources of language that Aristotle wished to assign to a purely secondary status. Such a speculation would no doubt have appealed to Gorgias, who could be expected to exploit the "logical" potential of tropes and figures in order to be able to respond to a multiplicity of occasions.

Aristotle's privileging of his own logical system and consequent devaluing of rhetorical "ornament" signals his resistance, then, to the sophistic view that associates meaning with the occasion of utterance. Aristotle in effect frees meaning from its dependence on occasion by a double strategy: first, he proposes that an ordered and coherent reality exists essentially unchanged from one occasion to the next; then, he argues that his own form of logic provides the means to replicate this structure in language. Instead of contriving a provisional, inherently one-sided representation of an ultimately un-

20. Ibid., p. 22.

graspable flux, the activity of invention can therefore discover enduring truth. The tropical possibilities of language, from which alternative logics might be fashioned, must take second place to Aristotle's own logic, a privileged repertoire of formal operations that eventually come to occupy the entire field of rhetorical persuasion. Aristotle's promotion of a plain style of logical clarity is thus a natural consequence of his hostility to sophistic skepticism, which held that truth emerges precariously from an uncertain encounter between mind and experience, neither of which possesses a determinable essence.

The contrast between Aristotelian and Gorgianic views on the heuristic possibilities of figurative language provides an early instance of a perennial controversy for which the twin rubric of "classicism" and "mannerism" may serve as a convenient label.[21] For the classicist, rhetorical tropes and figures are properly subordinate to the logical content of speech because language achieves its purpose in the preservation of a body of normative values. Aristotle would be an early exponent of classicism, since he argued that the style of a speech should not be obtrusive, that it must efface itself before the preexistent truth the orator would communicate. The mannerist, on the other hand, values the occasional over the traditional and restlessly experiments with the style of utterance in the hope of producing genuine novelty. Because of his skepticism concerning the possibility of knowledge, Gorgias will therefore practice a mannerist style, or one that makes ample use of rhetorical ornament. Since final knowledge is precluded for the sophist, content has to serve style.

But so long as their relationship is understood as one of surface to depth or accident to essence, it would be misleading to rely on the traditional opposition of "style" and "content." From Gorgias's point of view, "content" loses its customary meaning because this word usually implies some stable and permanent referent—precisely,

21. There is a useful discussion of the classicist/mannerist controversy in E. R. Curtius, *European Literature and the Latin Middle Ages*, trans. W. Trask (New York: Pantheon Books, 1953).

"Reality"—a fixed and abiding extralinguistic referent whose existence Gorgias would deny. "Content," or the world, is for Gorgias eternally renewed at each occasion and becomes, therefore, but one more level of "style." In this sense, Aristotelian logic, deprived of its certain purchase on the truth, its claim to replicate a preexistent state of affairs, can be redefined, accordingly, as one among many stylistic possibilities available to an orator. The content of a speech will thus be nothing other than successive layerings of rhetorical ornament. To put it another way, ethical and pathetic proofs, the figurative language Aristotle regarded as mere ornament, can be understood to provide the basis for elaborating alternative varieties of logical proof.

By relying on the inspiration of the moment to insure that speech answers uniquely to its time, mannerism counsels a stylistic pluralism that refuses to privilege one style over all others. The aesthetic effect the mannerist orator seeks would be one of sublimity, a sudden shock of awe, wonder, terror, euphoria, transporting an audience beyond a traditional standpoint. Alternatively, the classicist would seek beauty, understood as the repetition of tradition or the imitation of a prior reality. Hence Aristotle's conception of metaphor as a movement among synonymous terms: a classicist, he inhabits a world of beautiful correspondences rather than sublime discontinuities. The aesthetic category of the sublime would in fact have no place in the context of Aristotle's philosophically oriented speculations on rhetoric. "Sublimity" arises from irrational novelty, while Aristotle would communicate a preexistent truth. Briefly, then, where Aristotle holds that the orator's object is beauty (or repetition), Gorgias maintains that the orator should seek sublimity (or difference).

Occasioned by the emergence of the genuinely unprecedented, the aesthetic effect of sublimity can be said to depend on the tragic contrariety of experience, the inability of any dogmatic scheme to contain all the possible meanings of the world. The epistemological correlative of sublimity would therefore be skeptical *ataraxia,* said to ensue whenever the skeptic goes beyond opposites, beyond adherence to purely relative distinctions, so that instead of viewing the present occasion as a repetition of tradition, its historically contingent character can erupt into consciousness.[22] In his treatise "On Not-Being," Gorgias seeks to induce such a state of skeptical *ataraxia*. That is, "On Not-Being" would force a recognition of the antithetical or relative nature of truth. It can be interpreted as a philosophic justification both for Gorgias's promotion of *kairos* as the principle determining rhetorical invention, and for sublimity as the aesthetic effect that oratory should produce.

An antimetaphysical metaphysic, "On Not-Being" is a parodic inversion of Parmenides' doctrine of Being. According to Parmenides, the possibility of certain knowledge depends on the existence of a realm of eternal, immutable Being not subject to Becoming, or the accidents of history. This realm of pure Being, the origin of both Plato's doctrine of forms and Aristotle's notion of substance, must be located, therefore, somewhere beyond the everyday world, the flux of appearances that is consequently denied the status of essential Being. When Plato remarks that the sophists "take refuge in the darkness of non-being," he means that the sophists forsake the truth of Being when they turn their investigations to empirical reality. Plato's charge is an accurate characterization of Gorgias's project in the treatise "On Not-Being." "On Not-Being" would have as one of its implications an affirmation of the phenomenal world as final reality, an affirmation of "Not-Being," the flux that always exceeds knowledge and that can be defined only negatively as that which Being is not.

In his treatise "On Not-Being," Gorgias moves to subvert the Parmenidean doctrine of Being in a three-step argument: Gorgias argues first that nothing exists; second, that even if anything did exist, it would not be knowable; finally, that even if reality were knowable, it could not be communicated. As he begins his refutation of Eleatic doctrine, Gorgias seeks to establish the fundamental ambivalence and ambiguity of reality by first conceding that Not-Being, as nonexistent, is *not*, that is, does not exist, but, and here Gorgias

22. It is worth noting in this context that for the Surrealists, "the *point sublime* is the place at once abstract and precise, where all the contraries are identified"; see Mary Ann Caws, *The Poetry of Dada and Surrealism* (Princeton: Princeton University Press, 1970), p. 15.

delivers his most telling reproach of Parmenides, insofar as Not-Being *is* nonexistent, it *is,* as the diacritical contrary whose presence (or existence) is necessary if the contrastive meaning "Being" is to emerge. As Gorgias puts it, Not-Being "insofar as it is understood as nonexistent . . . will not exist, but insofar as it *is* nonexistent it will on the other hand exist."[23] On the basis of this parodically intricate ratiocination, Gorgias then playfully reverses Parmenides when he argues, "If the nonexistent exists, the existent will not exist, for these are opposites to each other, and if existence is an attribute of the nonexistent, nonexistence will be an attribute of the existent." That is, if thought can posit the existence of eternal immutable Being only by making a contrast with Not-Being, understood to mean the provisional and contingent, then "Being" can claim no mirroring correspondence with ultimate reality; it will comprise, rather, only a relative or partial truth. Gorgias can consequently claim that Being, as a universal all-embracing term, does not in reality exist.

The treatise "On Not-Being" is based, of course, on the diacritical characterization of knowledge implied by the doctrine of the *dissoi logoi.* But a more striking standpoint from which to elaborate on the consequences of Gorgias's argument is provided by the Chinese Taoist philosophers Lao-tze and Chuang-tze. In important respects, the views of the two Taoists parallel the argument Gorgias makes in the treatise "On Not-Being" and can therefore serve as the basis for an appropriate analogy. The identity of opposites that Gorgias takes over from the doctrine of the *dissoi logoi,* for instance, is echoed by Chuang-tze when he writes, "The 'that' is produced by the 'this' and the 'this' is also caused by the 'that'. This is the theory of mutual production. . . . Is there really a distinction between 'that' and 'this'? Or is there really no distinction between 'that' and 'this'?"[24] Chuang-tze's predecessor Lao-tze makes a similar point: "For although all

23. "On Not-Being" survives in two slightly diverging accounts, one by Sextus Empiricus, the other attributed to Aristotle, "On Melissus, Xenophanes, and Gorgias." This and subsequent quotations are from the paraphrase by Sextus Empiricus, which can be found in Sprague, ed., *The Older Sophists,* pp. 30–67.

24. This translation of Chuang-tze comes from Wing-tsit Chan, *A Source Book in Chinese Philosophy* (Princeton: Princeton University Press, 1963), pp. 182–83.

creatures under heaven are the products of Being, Being itself is the product of Not-Being."[25] This notion of the insubstantial or fictive character of the known universe (insubstantial because it diacritically depends on the nonexistent) is equally prominent in the famous passage that begins Lao-tze's *Tao Te Ching:* "The Tao that can be told is not the eternal Tao. The name that can be named is not the eternal name. The nameless is the beginning of heaven and earth. The named is the mother of ten thousand things."[26] In other words, although the everyday world of knowable entities presupposes a founding category of Being—or stability and permanence as the prerequisite for knowledge—"Being" itself depends on "Not-Being," a term that can be interpreted in either of two ways.

On the one hand, "Not-Being" is the name for the primordial flux that thought must overlook if it would construct an enduring edifice of knowledge. That is, the Tao or primordial Way of the world cannot finally be grasped by discursive reasoning because the ground of Being is not linguistic in character and cannot be exhausted in a linguistic representation. Not-Being is a swarming continuum without the distinctions on which knowledge needs to fasten if it would found itself once and for all. In this sense, "Not-Being" refers not so much to the so-called flux of phenomena, which is after all only an *idea,* than it does to "just this." On the other hand, contrary to the naive materialist who unconsciously construes matter according to the model of thought, the Taoist recognizes that the activity of thinking is constitutive of its objects because meanings exist diacritically with reference to other meanings. The very possibility of regularity in nature, for instance, depends on the idea of primal anarchy and formlessness. As Being requires Not-Being, so "truth" comes from "lying," the "eternal" from the "temporal," "sameness" from "difference," and *nomos,* or the systematic certitude of knowledge, from *kairos,* understood to represent chance or the exceptional. Putting the two points of view together (so that ontology and epistemology become, as it were, fused), one might

25. Chang Chung-yuan, *Tao: A New Way of Thinking* (New York: Harper and Row, 1977), pp. ix, 102–3.

26. Lao-tze, *Tao Te Ching,* trans. Gia-fu Feng and J. English (London: Wildwood House, 1973).

conclude, then, that reality cannot be coaxed into the form of the statement "Being and not Not-Being." Reality, which does not obey the law of the excluded middle, cannot be made to conform to an "either/or" structure. Rather, as the Greek tragedians recognized, it should be thought of as "both/and," both Being and Not-Being, chaos and cosmos, or as Joyce terms it, "chaosmos." Being and Not-Being are equally existent nonexistent constituents of reality.

The skeptical *ataraxia* ideally produced by the treatise "On Not-Being" would result, then, in a mental attitude resistant to dogmatic self-enclosure, abiding neither in Being nor in Not-Being. The mind would return, in effect, to the center that is no-center because it comprises both sameness and difference, rest and movement, the center of the "here now," the absolute present, most ineffable because most concrete of facts; a center that is never available as a simultaneity and can never become, therefore, the object of any mode of consciousness. Chuang-tze points out, "Heaven and Earth and I live together, and all things and I are One. Since we are all One, how can we express the One? If we express the One, our expression and the One become two."[27] Even the intention to be intentionless, to attend to reality as just itself, constitutes a relative point of view, or approaches experience according to a preconception. Just as there can be no final conceptual understanding of reality, so there can be no unmediated intuitive apprehension of the essence of things. Mental experience of no matter what sort is inflected from a relative point of view. Accordingly, avoiding dogmatism on the one hand and quietism on the other, the mind must resign itself to unending improvisation and experiment.[28]

In the subsequent unfolding of the treatise "On Not-Being," Gorgias denies the representational efficacy of language in an argument to the effect that ultimate truth can be neither known nor communicated. If tragic contrariety is indeed the leading charac-

27. Quoted in Chang, p. xvi.

28. Colin A. Ronan explores the relationship between Taoism and experimental thought, in *The Shorter Science and Civilization in China* (Cambridge: Cambridge University Press, 1978), vol. 1, an abridged version of Joseph Needham's original text. Ronan remarks there that Taoism "was the only system of mysticism the world has ever seen that was not profoundly anti-scientific" (p. 85).

teristic of reality, then knowledge cannot stand in a relationship of simple correspondence with the world. Thought's inability to impose a lasting unity on the multiplicity of experience means that thinking and speaking must finally be recognized for just what they are, as palpable bodily experiences rather than transparent instruments of representation: "Just as objects of sight are said to be visible for the reason that they are seen, . . . so, too, things considered in the mind will exist even if they should not be seen . . . , because they are perceived by their own criterion." The activity of thinking is proof only of the existence of thought. Thinking transpires in a mental space unillumined by any external criterion that would enable a thought corresponding with reality to be distinguished from a fantasy. The world of thought comprises its own reality whose correspondence with the world at large is secured by no dispensation of representational truth coeval with the onset of thought. Gorgias would thus dispute the Eleatic claim that metaphysical speculation, on the basis in effect of visionary revelation, can discover essential Being. According to Gorgias, the truth value of a thought derives from its ability to persuade, from the sensuous properties of speech. Discovery of the truth involves equally its fabrication, the ultimately arbitrary imposition of form upon the teeming variety of the world. The meanings we attribute to reality are not naturally resident in the world but depend, finally, on rhetorical artifice, on the persuasive force of the speaker, for "that by which we reveal is *logos,* [and] *logos* is not substances and existing things. Therefore, we do not reveal existing things to our neighbors, but *logos,* which is something other than substances." The authority of the truth derives from the particular context of rhetorical persuasion, from the always provisional persuasive artifice of the speaker, who plunges into the swarming continuum of experience in order to retrieve from it a knowable world that is, in the last analysis, purely a verbal figment able to persist only so long as thinking goes on.

Gorgias's identification of truth with rhetorical artifice carries with it the implication that deception is intrinsic to any act of persuasion. In his "Encomium on Helen," Gorgias develops this thesis in the course of an argument seeking to remove blame from the reputation of Helen of Troy. According to Gorgias, if Helen was indeed "by words seduced," or persuaded by Paris to desert Menelaus, then her name deserves to be rehabilitated because "the persuader, like a constrainer, does the wrong, and the persuaded, like the constrained, in speech is wrongly charged." Helen should not be held responsible for acceding to Paris because persuasion of no matter what sort is based on deception. As Gorgias puts it, "All who have and do persuade people of things do so by molding a false argument," for in the absence of certain knowledge of past, present, and future, human beings are forced to "take opinion as counselor to their soul, but . . . opinion is slippery and insecure." Since all persuasion depends on opinion, on a semblance of the real rather than on reality itself, Helen is no more culpable than any other audience persuaded—that is, deceived or seduced—by the rhetorical artifice of an orator.

As he reviews respectively aesthetic, ethical, and epistemological functions of language, Gorgias moves to supplant the traditional characterization of speech as instrument for conveying the truth. Speech, he says, "is a powerful lord, which by means of the finest and most invisible body effects the divinest works: it can stop fear and banish grief and create joy and nurture pity." Emphasizing the pathetic impact of speech, Gorgias argues that just as poetry can create "fearful shuddering and tearful pity and grievous longing" in an audience, so sacred incantations "are bearers of pleasure and banishers of pain, for, merging with opinion in the soul, the power of incantation is wont to beguile and persuade it and alter it by witchcraft." Gorgias thus identifies the aesthetic and ethical functions of language with divine possession and magic, with incantation that produces Beauty and the Good according to no rational determination. Next turning to the epistemological import of speech, Gorgias remarks that when the natural philosopher scientifically explains the phenomenal world, so that "what is incredible and unclear seem true to the eyes of opinion," or when the "swiftness of thought" of the philosopher engaged in verbal dispute makes "belief in an opinion subject to easy change," the effect of speech "upon the soul is comparable to the power of drugs over bodies." In this sense, the "Encomium on Helen" can be said to conclude by characterizing the orator as a *psychagôgos*, a leader of souls whose oratory deceives, seduces, bewitches, or drugs an audience.[29] Gorgias intends to stress that language is as much productive of reality, or persuasive on the basis of its own intrinsic resources, as it is a vehicle for representation.

But the orator is not simply a seducer who imposes subjective will by means of deception, for the "logic of seduction," as it were, imposes itself equally on the seducer and the seduced.[30] That is, in order to achieve success, the orator as seducer must be "seduced" in turn by the occasion of speech. Persuasion depends on self-effacement, on acceding to the terms laid down by the circumstances confronting the speaker. Chuang-tze makes a characteristically radical statement of this requirement when he argues for an activity of interpreting the world that would ensue in the disappearance of the self as it becomes identified with whatever contingent manifestation passes before its attending gaze: "Exercise fully what you have received from Nature without any subjective viewpoint. In one word, be absolutely vacuous."[31] Lacking a fixed identity, the orator's self would thus become identical with its present performance. Concerning this ideal absence of a self-nature, William James remarks, "If we could say in English 'it thinks', as we could say 'it rains' or 'it blows', we should be stating the fact most simply and with the minimum of assumption."[32] Lao-tze's formulation in the *Tao Te Ching*, on the other hand, perhaps better preserves the paradoxical character of this ideal interaction of the self with its environment: "It is because [the sage] has no selfishness that he can make his selfishness succeed."[33] The seducer who is seduced by the occasion

29. Kennedy observes that for Gorgias the orator is "a *psychagôgos*, like a poet, a leader of souls through a kind of incantation" (p. 62).

30. For an interesting discussion of this issue, see Mario Perniola, "Logique de la séduction," trans. Thierry O. Séchan, in *Traverses* 18 (February 1980): 2–9.

31. In Chan, p. 207.

32. James, pp. 224–25; for a finely nuanced account of the self as an "*act*, as a composition, a performance, a 'making,'" see Richard Poirier, *The Performing Self: Compositions and Decompositions in the Languages of Contemporary Life* (New York: Oxford University Press, 1971), p. 90.

33. There are many translations of the *Tao Te Ching;* this one is quoted in Raymond Dawson, ed., *The Legacy of China* (Oxford: Oxford University Press, 1964), p. 43.

of seduction can be taken as the emblem for an ideal dynamic between a principle of intentionality (or the self that would repeat itself in the world) and a principle of spontaneity (or the unforeseen opportunity of the immediate occasion).

The career of the sophist would therefore unfold as a loosely connected sequence of encounters with the world, something like the episodic structure of the tale told by Odysseus in the *Odyssey,* a series of incidents arranged finally according to formal and aesthetic rather than chronological or causal criteria, or like Ovid's nonteleological view of history in the *Metamorphoses,* in which history is deployed as a random succession of moments of transformation. The *ethos* of the sophist would be a fluid one, implying an occasional or context-specific stance toward experience. Instead of exhibiting an achieved or final version of the self, in sophistic *epideixis* the sophist would enact an interminable interpretive activity. The process of making sense can be understood to imply a willingness to risk the history of the self in a mutual transformation of subject and object; or as Gertrude Stein would say, a willingness to begin again and again, and so undertake an endless task of adaptive metamorphosis.

Perhaps the crucial condition imposed on Gorgias when he wrote the treatise "On Not-Being" and the "Encomium on Helen" was the need to adapt his arguments to a requirement that in fact attaches to any act of communication. For communication to take place, the historical specificity of the present must momentarily be overlooked so that the fiction of shared understanding can emerge, the fiction of a stable world beyond appearances, accessible to thought, and able to be communicated. In a sense, Gorgias had to forget contradiction in order to communicate "contradiction," since true mimesis of the contradictory character of reality would devolve into incomprehensible dissonance. The dilemma confronting Gorgias can be put as follows: how can one communicate the "unheard of" in such a way that it is understandable but not immediately assimilable to the *doxa?* According to Gorgias, this problem has already been resolved in tragic drama, "a deception in which the deceiver is more justly esteemed than the nondeceiver and the deceived is wiser than the undeceived." That is, tragedy in effect musicalizes dissonance, endowing the contradictory, antithetical character of the truth with a beautiful or comprehensible form.

In *The Birth of Tragedy,* Nietzsche similarly characterizes tragic drama as a true deception, a lie that tells the truth, an ideal intersection, in his vocabulary, of the "Dionysian" with the "Apollonian." For Nietzsche, the Dionysian is "the desire to tear asunder the veil of Maya, to sink back into the original oneness of nature," a state of ecstasy that seeks to merge with "the constant proliferation of forms pushing into life . . . the extravagant fecundity of the world will." Were it not for the Dionysian principle of irrational exuberance, Apollonian rationality and purposiveness would freeze, Nietzsche says, into Egyptian rigidity. But if the Dionysian were permanently to displace the Apollonian, it would suicidally live out the strife of opposites in nature. Its exorbitant, chaotic vitality would cause it to go to war with itself. Therefore, the original oneness, "the ground of Being, ever suffering and contradictory, time and again has need of rapt vision and delightful illusion to redeem itself." Human being, the very "incarnation of dissonance," to be able to endure life needs "a marvelous illusion to cover it over with a veil of beauty," requires "countless illusions of fair semblance" in order to "whet our appetite for the next moment."[34]

The truth, in other words, needs to be dissembled, as Gertrude Stein suggests when she characterizes her project at one point in her career in this way: "There must be distribution and equilibration, there must be time that is distributed and equilibrated."[35] Stein thus recognized that the experience of radical sublimity has to be *harmonized,* that she had to produce "a harmony in hesitation" if she was to communicate her meaning. That is, the experience of sublimity must be musicalized, must appear in the guise of the beautiful, must dissemble itself if assent is to be secured for difference or "suchness," for "a violent kind of delightfulness" as a positive virtue. In order to install the sublime as a truth within the mind, repetition,

34. Nietzsche, *The Birth of Tragedy and the Genealogy of Morals,* trans. F. Golffing (Garden City, N.Y.: Doubleday, 1956), pp. 27, 102, 32, 145.

35. Stein, "Composition as Explanation," in *Selected Writings,* p. 522.

copiousness, a beautiful dissembling of sublimity is needed if the singular and exceptional is to be accepted as the routine. In this sense, *Gorgias must become Aristotle.* If Aristotle's philosophically oriented rhetoric makes no provision for the emergence of radical novelty, Gorgias's reliance on the irrational spontaneity of the immediate occasion risks incomprehensibility if not incoherence. Gorgias must therefore interpret the world, at least temporarily, in a manner consistent with a particular logical system. And he must resort to tradition as well, to a repertoire of *topoi* or commonplace beliefs, so that the most fitting term to describe Gorgias's rhetorical practice would be *exergasia* (otherwise and appropriately known as "gorgious"), an effect of variety achieved nevertheless by means of repetition, by varying a theme: a beautiful semblance of sublimity.

Gorgias's rhetorical practice, which would induce an audience to accept no-criterion as the criterion or normative value regulating conduct, thus obeys a contradictory injunction. The desire that thought should continually innovate, so that rather than simply repeating, it would always posit alternatives to tradition, is accompanied by the recognition that thinking must become complicit with tradition if it would communicate with an audience. Gorgias is able to maintain this tension between spontaneity and tradition, too easily resolved in favor of tradition, by means of his literary extravagance. Gorgias's prose spectacle tells the truth about deception by emphasizing that meaning is only a figment conjured up by the tropes and figures of language. His mannerist style forces the recognition that sublimity provides only a point of departure for a practice of intentional repetition, a beautiful but deceptive insistence on a provisional start. That is, to seize the time, a shape or form of coherence must be imposed upon it. But any coherence deceptively narrows the tragic contrariety of experience. Thought must therefore always keep on the move, lest it freeze into the rigidity of complacent self-delusion. In order to promote such mobility, Gorgias deploys his extravagant style with the intention that the music of language will seduce its audience away from the music of the spheres, or from belief in a reality preexistent to thought. Gorgias's style would even carry with it the implication that the mind deceives itself at the very moment that it postulates the delusory character of all knowledge. Gorgias would disclose that "tragic contrariety," in the last analysis, is only a figure of speech.

Gorgias's rhetorical practice remains forever suspended, then, between the singular and the universal, singular in that it embraces only a relative point of view, universal in that it proclaims the relativity of any standpoint. This dilemma has important consequences for the present essay, the title of which is meant to serve notice of the unresolvable tension between spontaneity and tradition in the midst of which the will-to-invent must intervene. *Kaironomia* cannot claim for itself a privileged finality. It will attempt to resist the inclination to dogmatically elaborate an argument according to a closed conceptual system. But in so doing, it will in fact practice a double movement: while denying itself the imaginary purchase on Being secured by an interpretive procedure that quests after a final meaning, it will nevertheless retain the intention to "speak the truth"; but a "truth" that is always approximate, always provisional, always a dissembling. Thus, if *Kaironomia's* argument is admitted to be an improvisation, a contingent and provisional response to a particular occasion, then it can promise no "true" knowledge. But if the implication of this argument is precisely that no "true" knowledge exists, then the argument foresees or "knows" its own historically contingent character and, in that sense, rises above its particular situation to a transtemporal realm. If *Kaironomia* is only a kaironomia, then what it says is not truth but lie or mere opinion. Alternatively, if *Kaironomia* announces that it is an instance of kaironomia, then it is telling the simple truth. The argument must fall prey, then, to the Paradox of the Liar.[36] Concluding with an avowal of its own inconclusiveness, *Kaironomia* will slide undecidably between the promise of transtemporal knowledge and an admission of its inability to transcend the historical determinations of its own occasion. At once a theory and a practice, it remains caught in an inevitable double bind between an intention to "speak the truth" about the endlessness of interpretation and an awareness that

36. The Paradox of the Liar: If I am a liar and I say "I am a liar," then I am not a liar; but anyone who is not a liar who says "I am a liar" is a liar.

every "truth" (including the one that would posit "no-truth") is historically determined.

This essay will complete the following transit: in Part 2, the modern recurrence of a Gorgianic conception of Being, Knowledge, and Communication will provide an opportunity to amplify the theory of rhetorical invention sketched in this introductory section; Part 3 will consist of a practical demonstration of the theory developed in Part 2; while Part 4 will offer a concluding restatement of the main points at issue. *Kaironomia* will follow, then, a customary itinerary of Introduction, Argument, Proof, and Conclusion. But however much it may conform to the desire for linear movement toward a conclusive terminus, the movement of this essay will necessarily also entail a series of returns, as it were, to its own starting point. On the one hand, a philosophic view of rhetorical invention connecting meaning with immediate circumstance, on the other, *Kaironomia* must insist on its own provisional character. Returning to itself as occasion of rhetorical invention even as it explores the nature of speculative thought, the present essay can be said to spiral away from its point of departure.

PART TWO

The "Reflexive" Is (Not) the "Middle Voice"

The nature of infinity is this: That everything has its
Own Vortex, and when once a traveller thro' Eternity
Has pass'd that Vortex, he percieves it roll backward behind
His path, into a globe itself unfolding like a sun,
Or like a moon, or like a universe of starry majesty,
While he keeps onwards in his wondrous journey on the earth,
Or like a human form, a friend with whom he liv'd benevolent.
As the eye of man views both the east & west encompassing
Its vortex, and the north and south with all their starry host,
Also the rising sun & setting moon he views surrounding
His corn-fields and his valleys of five hundred acres square,
Thus is the earth one infinite plane, and not as apparent
To the weak traveller confin'd beneath the moony shade.
Thus is the heaven a vortex pass'd already, and the earth
A vortex not yet pass'd by the traveller thro' Eternity.

—William Blake, *Milton*

I

In his 1915 essay "Instincts and Their Vicissitudes," Freud proposes that the derivation of sadomasochism he presents there be understood with respect to three grammatical categories. Sadistic mastery, an essentially transitive relationship to the world, corresponds, Freud says, to the active voice, while the pleasure the masochist takes in being dominated corresponds to the passive voice. And in an intermediate step, where the subject pursues a program of self-mastery, "the active voice is changed, not into the passive, but into the reflexive middle voice."[1] The various positions available to the self as it negotiates the vicissitudes of sadomasochism—either as master of the world, master of itself, or mastered by the world—can thus be correlated with the active, reflexive middle, and passive voices, each a transformation of the other. Freud suggests, then, that the metamorphosis of desire obeys a formal necessity. Instinctual energy is orchestrated, finally, according to a grammar of desire.

As a complementary implication to this first point of view, Freud's argument can also be understood to claim an erotic basis for grammar. That is, active, passive, and reflexive middle voices should not be construed simply as formal devices, conventions for making sense of the world. Beyond their instrumental utility, active, passive, and reflexive middle have an erotic purpose to fulfill. They represent strategies, in and of themselves, for producing pleasure. According to this second implication of Freud's argument, the syntax with which the self frames its knowledge of the world is deployed from

1. Freud, "Instincts and Their Vicissitudes," in *The Standard Edition of the Complete Psychological Works of Sigmund Freud* (hereafter cited as *SE*), ed. James Strachey (London: Hogarth Press, 1953–74), 14:128.

the beginning in the service of an erotic motive that is essentially indifferent to reality.[2] In other words, a self-delighting expense of desire is intrinsic to the process of coming to know the world. Freud's correlation of sadomasochism with grammatical voice thus has the effect of expanding the field of desire, so that apparently nonsexual and purely functional grammatical categories come to reside within the domain of the erotic.

Desire can be described, Freud says, as a psychic movement from a condition of tension to a pleasurable release. According to the psychoanalytic account of the genesis of desire, desire emerges within the organism as a stimulus seeking its own abolition in a pleasurable spending of instinctual energy. Every desire demands satisfaction. Freud defines the nervous system as "an apparatus having the function of getting rid of the stimuli that reach it, or of reducing them to the lowest possible level: an apparatus which would even, if it were feasible, maintain itself in an altogether unstimulated condition."[3] Erotic pleasure thus ensues from a reduction of the desiring tension provoked in the organism in the course of its interaction with its environment. A destabilizing intrusion into the psychic equilibrium of the self, the emergence of desire is always experienced, at least in part, as an irritation, an irritation that spurs the movement of desire toward a condition of nondesiring stillness.

Sadism and masochism provide two strategies for circumventing the disruptive energy of desire. Each seeks, by channeling the flow of instinctual energy along a routine itinerary, to limit the proliferation of desire. Sadism and masochism therefore represent a conservative tendency within the organism that would sooner discharge in-

2. In *Civilization and Its Discontents,* Freud notes that an erotic motive persistently encroaches on the self-preserving instrumental behavior of the organism. The ability to distinguish subject from object, for instance, is inflected from the beginning by the desire for pleasure: "A tendency arises [in infancy] to separate from the ego everything that can become a source of . . . unpleasure, to throw it outside and to create a pure pleasure-ego which is confronted by a strange and threatening 'outside.' . . . Some of the things that one is unwilling to give up, because they give pleasure, are nevertheless not ego but object; and some sufferings that one seeks to expel turn out to be inseparable from the ego in virtue of their internal origin" (*SE,* 21:67).

3. Freud, "Instincts and Their Vicissitudes," *SE,* 14:120.

stinctual energy according to a habitual mode of satisfaction rather than by improvising a new one.[4] Both sadist and masochist would render the achievement of erotic satisfaction routine, and so release the self from the burden of invention. Thus, sadistic arousal depends on the unresisting compliance of the masochist, whose satisfaction derives in turn from vicarious participation in the autonomy, stability, and mastery attributed to the sadist. Both aspire to a condition of rest beyond the irritating intrusions of desire, a final version of the self beyond the flux of history.

This aspiration for a permanent state of erotic satisfaction can be understood to represent desire's fundamental desire, as it were, not to be desiring. Obedient to a suicidal impulse, every desire intends its own death. What Freud will later refer to as the "death instinct" is in fact nothing other than the desire to reestablish a condition fantasized as anterior to the disorienting excitement produced in the organism when it comes into contact with its surroundings. The "death instinct" is desire in its primary form, the point at which sadism and masochism coincide. As Freud remarks in his 1924 essay "The Economic Problem in Masochism," the death instinct "which is operative in the organism—primal sadism—is identical with masochism."[5] The underlying identity of sadism and masochism can be formulated in this way: the sadistic project for the exercise of absolute control would amount to a definitive incorporation of the world in which the line dividing subject from object would be erased. The sadistic exercise of absolute power would thus simultaneously be experienced as masochistic pleasure. Mastery *of* the world becomes mastery *by* the world. Sadistic and masochistic eroticism depend on the prospect of a definitive moment of unambiguous consuming intensity free from the irritating lack of desire. There is only a dif-

4. In this connection, Freud remarks that "the difficulty of cultural development" can be traced "to the inertia of the libido, to its disinclination to give up an old position for a new one"; see *Civilization and Its Discontents, SE,* 21:108.

5. Freud, "The Economic Problem in Masochism," *SE,* 19:164; two works important for this discussion of the nature of desire are Leo Bersani, *Baudelaire and Freud* (Berkeley and Los Angeles: University of California Press, 1977); and Jean Laplanche, *Life and Death in Psychoanalysis,* trans. Jeffrey Mehlman (Baltimore: Johns Hopkins University Press, 1976).

ference of emphasis, then, between the sadist's dream of devouring the world, a feat to be accomplished by pure naked aggression, and the masochist's dream of being devoured by the world, achieved when the masochist is finally fixed in the regard of an all-powerful other.

But to the extent that sadism and masochism remain distinguishable as forms of behavior, the respective strategies they provide for eliminating desire must ultimately prove ineffectual. Sadist and masochist are caught in a logic of mutual implication. As sadistic pleasure depends on the presence of a submissive partner, so masochistic pleasure requires a dominant one. Each remains forever vulnerable, therefore, to an intrusive desire for its symmetrical opposite. Only in the coincidence of sadism with masochism—where mastery of the world would be equivalent to self-mastery—could a lasting condition of autonomy and stability be achieved. According to Freud, the intermediate step in his derivation of sadomasochism—grammatically, the "reflexive middle voice"—comprises such a coincidence. Embodying the dual character of the death instinct, at once active and passive, the reflexive middle voice can be characterized as a culminating strategy for the immobilization of desire. It provides the means, as it were, for finally freeing the self from the "shock of the new," from the burden of inventing an original response to the world on the spur of the moment. By seizing hold of itself as definitive object of desire, the subject would at last be able to freeze the meaning of the world in a moment of lasting comprehension, and so remove the possibility of further desiring movement.

Freud suggests, however, that even this strategy for putting an end to the movement of desire represents, finally, a futile gesture. The subject's efforts at self-enclosure inevitably will be disrupted, he says, because instinctual stimuli "maintain an incessant and unavoidable afflux of stimulation."[6] The unconscious is an infinite reservoir of desires seeking satisfaction. The satisfaction of desire is therefore an infinite task, for the restless energy propelling the subject's desiring movement can never be extinguished. Desire always already "knows," in effect, that no plenitude imaginable will ever

6. Freud, "Instincts and Their Vicissitudes," *SE*, 14:120.

prove commensurate to its own exuberance, abundance, and excess. The unending emergence of desire within the organism, continually disrupting the psychic equilibrium of the self, is a permanent confirmation of the fact that the self, as Freud puts it, is not the master in its own house. No self is ever adequate to the protein diversity of its own experience. The disintegrating agency of desire ceaselessly undermines the subject's attempt to attribute finality to its present stage of realization. Sadomasochistic pleasure can never be achieved once and for all. As desire is continually renascent within the body, so dominance and submission must endlessly be renewed. Instead of a permanent condition of rest, the sadomasochistic self can enjoy at most the unending repetition of a temporary satisfaction.

The attempt to arrest the process of history at a moment of plenitude thus devolves into inflexible self-repetition, merely a rigidification of the self rather than a genuinely achieved conclusive state. Freud's description of the nature of desire pessimistically implies a tendency, as it were, to philosophic dogmatism as part of the erotic constitution of the individual. But the analogy Freud makes between the stages of sadomasochism and the categories of grammatical voice may provide an alternative to this ultimately futile quest for permanent equilibrium. There is a fortunate confusion in Freud's reference to the "reflexive middle voice" as a way of characterizing the project for a definitive immobilization of both subject and object of desire. As Jean Laplanche observes in *Life and Death in Psychoanalysis,* the so-called reflexive middle voice confuses two distinct grammatical forms: "The reflexive form distinguishes more clearly the subject and the object of the action, allowing the fantasized exchanges of position to take place [when the sadist identifies with the intensity of the masochist's suffering, and the masochist fantasmatically appropriates the sadist's illusion of control]. In the 'middle' form, the terms of the fantasy remain in something of a state of coalescence."[7] Although the two grammatical forms share in a recursive movement, in the reflexive form the self is fixed in its own regard as it seeks to reproduce itself identically, while in the middle form the self goes out into the world to become what it is

7. Laplanche, *Life and Death in Psychoanalysis,* p. 143.

not, so that even as it repeats itself, it makes a difference. The "coalescence" of subject and object transpiring in the middle voice thus describes a destabilization of the self that genuinely changes the subject, reconstituting the self on a new basis.

The middle voice provides an alternative, then, to the sadomasochistic desire to establish a final version of the self. Forgoing the reassuring illusion of finality, in the middle voice the self would serve only as a provisionally privileged center of pleasure. Such a self would lead a fugitive existence, always on the move from one newly constituted version of itself to another. The middle voice suggests not a fixed and abiding selfhood but a sequence of discontinuous partial selves, or the self as a historical process. Transformed even as it transforms the world, the self would continually improvise a response to its experience on the basis of fresh intrusions of desire from outside itself. Rather than seeking to limit the limitless productivity of desire, the middle voice would promote an activity of endless desiring metamorphosis.

In this sense, the "middle voice" can be understood as a possibility of movement among active, passive, and reflexive forms. To put it another way, as active and passive are contained within the reflexive, so the reflexive would comprise a relative moment within the middle voice. The middle voice would be a version of the reflexive able to recognize that perfect mastery is finally an illusion and, therefore, that invention must constantly be renewed if the self is to successfully adapt to its present situation. Where the masochist believes that everything that needs to be said has already been said, and the sadist impatiently looks forward to the day when there will be nothing more left to be said, in the middle the self would programmatically reject the prospect of the end of knowledge. But even though it rejects both passive acquiescence to tradition and the aggressive pursuit of final mastery over the self's surroundings, the middle voice would be compounded equally, as it were, of either strategy. That is, it implies a practive of speculative thought in which a receptiveness to tradition is able to coexist with an active intention to achieve mastery precisely because the culminating reflexive dream of a perfect illumination of subject and object of knowledge has been converted in effect into a continually destabilizing skeptical self-regard. In the middle voice, then, active, passive, and reflexive forms would no longer operate in the service of the desire for definitive erotic satisfaction. As relative moments within the middle, they would instead provide *topoi* or points of departure for rhetorical invention, strategies not for remaining the same but for becoming other.

The very existence of a grammatical category so protean and indeterminate in nature as the middle voice is necessarily a precarious one. The middle voice in fact disappeared as an identifiable linguistic form during antiquity when, as it were, a "hardening" of grammatical categories led to its replacement by the fixed opposition of the active/passive dichotomy.[8] In his 1966 essay "To Write: Intransitive Verb?" Roland Barthes argues that since it is no longer available as an expressive possibility in ordinary usage, the middle voice should therefore be understood as a historical antecedent or prototype for the mode of discourse to which contemporary thought aspires. The task, he says, is "to rethink the lost category and to take it as a metaphorical model . . . to reclaim it by raising it to the level

8. Jean-Pierre Vernant has remarked that the disappearance of the middle voice coincides with the gradual emergence during the Classical period "of a vocabulary of the will . . . precisely the idea of the human subject as agent, source of actions," the idea, in other words, that the subject authors rather than participates in its experience; in Macksey and Donato, eds., *The Structuralist Controversy*, p. 152. In his *Problems in General Linguistics*, trans. M. E. Meek (Coral Gables, Fla.: University of Miami Press, 1971). Emile Benveniste notes that originally in the Indo-European languages there was an opposition between the active and middle voices: "In the active, the verbs denote a process that is accomplished outside the subject. In the middle . . . the verb indicates a process centering in the subject. . . . The subject is the center as well as the agent of the process, he achieves something which is achieved in him—being born, sleeping, lying, imagining, growing, etc. He is indeed inside the process of which he is the agent" (pp. 148–49). According to Benveniste, the middle voice was subsequently transformed into the passive, thereby changing the meaning of the opposition. That is, by concentrating merely on the activity or passivity of a subject whose identity remains essentially unaffected by the events in which it is implicated, the active/passive opposition encourages the view that individual identity precedes rather than emerges from a history of encounters with the world. The reflexive voice that Freud confused with the middle depends on this later distinction.

of discourse."[9] The loss at the grammatical level of the middle voice can thus be compensated for at the rhetorical level in an activity of invention that continually recurs to its own occasion, a practice of speculative thought in which subject and object are engaged in an endless process of mutual transformation. Always ready to revise the meaning it atrributes to the world, this renewed form of the middle voice would affirm the processual character of reality. It would be consistent with this description of reality from Ernest Fenollosa's *The Chinese Written Character as a Medium for Poetry*—"A true noun, an isolated thing, does not exist in nature. Things are only the terminal points, or rather the meeting points, of actions, cross-sections cut through actions, snapshots. Neither can a pure verb, an abstract motion, be possible in nature. The eye sees noun and verb as one: things in motion, motion in things."[10] Or as James Joyce puts it in a passage from *Finnegans Wake* whose rhetorical practice already in effect revives the middle voice: "The man from Saint Yves may have been (one is reluctant to use the passive voiced) may be been as much sinned against as sinning, for if we look at it verbally perhaps there is no true noun in active nature where every bally being—please read this mufto—is becoming in its owntown eyeballs."[11]

Barthes draws on several concepts from linguistic theory in order to characterize the opportunity for speculative invention suggested by the middle voice. With regard to temporality, he argues that discourse must always be contemporary with itself, alert and able to adapt to the present occasion. The writer must therefore "learn to distinguish between the present of the speaker, which is grounded in a psychological fullness, and the present of what is spoken, which is mobile and in which the event and the writing become absolutely coincidental." And with regard to person, Barthes contends that the

9. Roland Barthes, "To Write: Intransitive Verb?" in Macksey and Donato, eds., *The Structuralist Controversy;* actually, Barthes makes this statement in the conversation appended to his essay, p. 152.

10. Ernest Fenollosa, *The Chinese Written Character as a Medium for Poetry,* excerpted by Donald Allen, ed., in *The Poetics of the New American Poetry* (New York: Grove Press, 1973), p. 17.

11. James Joyce, *Finnegans Wake,* 3d ed. (1964; London: Faber and Faber, 1975), p. 523.

writing subject must always be in the act of creating itself anew: "When a narrator recounts what has happened to him, the 'I' who recounts is no longer the same 'I' as the one that is recounted. In other words . . . the 'I' of discourse can no longer be a place where a previously stored up person is innocently restored." To assert the contrary, namely that there is an essential self persisting through time, is to commit "bad faith," Barthes says, by making "literary form simply the expression of an interiority constituted previous to and outside language."[12]

When Barthes warns against privileging the author as the originating source of utterance, he intends to undermine the idea of a noncontingent or essential self. For Barthes the notion of the self as a discrete and autonomous entity represents the most stubborn form of dogmatism. According to this view, the self is always dependent, even when it refuses to acknowledge the fact, on "the voice of the Other." Psychoanalytically, the voice of the Other can be understood as displacement, an aleatory movement of desire toward as yet undisclosed destinations. Tropically, it corresponds to metonymy, a movement from one element to another on no more motivated a basis than simple contiguity. By continually displacing the controlling center within the subject, the disruptive agency of metonymy enables a theoretically endless mutation of thought. In fact, it forbids our regarding invention as dependent, either objectively or subjectively, on a preexistent plenitude. Neither the world represented in a text nor its author can be presupposed as originating presences. The subject and object of discourse must be renewed at each occasion. They can never be more than fluid and relative moments in the flux of signification. We are enjoined, therefore, to transcend the ego's tendency to attach itself to an illusory plenitude in the hope of having its dream confirmed of truly being the author of its experience. The disruptive agency of metonymy suggests an aesthetic of surfaces, of abrupt shifts and juxtapositions, a movement of thought

12. Roland Barthes, "To Write: Intransitive Verb?" pp. 138, 140, 141; in this connection, compare Gertrude Stein's remark in *Everybody's Autobiography* (New York: Random House, 1937; p. 102) that "neither character nor identity are necessary to him who meditates." For Stein, to accede to the notion of a self fully present to itself would be to fall prey to memory, which is repetition and confusion.

in which desiring energy is invested in the present. Instead of the reflexive dream of purely habitual uninventive behavior, it implies a rhetorical practice in which the speaker's intention emerges in response to the particular situation.[13]

For a rhetorical practice truly to be alert to its own occasion, it would have to be suspicious of received cultural forms on the grounds of dogmatism and ideological complicity. A renewed version of the middle voice could be characterized in this sense as a resurgence of philosophic skepticism. In an essay entitled "Semiology and Rhetoric," Paul de Man has usefully touched on the issue of skepticism in a comparison implicitly drawn between skeptical *ataraxia* and the effect of discovering an undecidable structure in a literary text. In his commentary on a passage from Proust, de Man observes that "individual metaphors" (or moments when a fullness of meaning prevails) "are shown to be subordinate figures in a general clause whose syntax is metonymic," merely a stringing together of unrelated elements rather than a genuine unity. "But this metonymic clause has as its subject a voice whose relationship to this clause is again metaphorical"—voice as the token of a psychic totality anterior to the moment of utterance—and this "subject-metaphor" is vulnerable in turn, de Man says, to a similar undoing. The undecidability produced by the alternation of epistemologically incompatible tropes results in "an anxiety (or bliss, depending on one's momentary mood or individual temperament) of ignorance."[14] The "bliss" to which de Man

13. Barthes describes this purely circumstantial activity of invention in characteristically utopian terms in his 1968 essay "The Death of the Author": "The modern scriptor is born simultaneously with the text, is in no way equipped with a being preceding or exceeding the writing, is not the subject with the book as predicate; there is no other time than that of the enunciation and every text is eternally written *here and now*. The fact is (or, it follows) that *writing* can no longer designate an operation of recording, notation, representation, 'depiction' (as the Classics would say); rather, it designates exactly what linguists, referring to Oxford philosophy, call a performative, a rare verbal form (exclusively given in the first person and in the present tense) in which the enunciation has no other content (contains no other propositions) than the act by which it is uttered—something like the *I declare* of kings or the *I sing* of very ancient poets."

See Barthes, *Image-Music-Text,* trans. Stephen Heath (New York: Hill and Wang, 1977), pp. 145–46.

14. Paul de Man, "Semiology and Rhetoric," in Josué V. Harari, ed., *Textual Strategies: Perspectives in Post-Structuralist Criticism* (Ithaca: Cornell University Press, 1979), p. 140.

refers is, of course, the bliss of *ataraxia* that results whenever the skeptic goes beyond adherence to purely relative distinctions.

But the term *skepticism* is perhaps misleading, as de Man has noted in another context: "I prefer the term 'irony' to that of 'skepticism.' . . . Irony goes beyond scepticism in that it puts the notion of the self in question. Scepticism is an attitude of consciousness while irony is a purely linguistic trope."[15] "Irony" is preferable to "scepticism" because irony undermines the installation of the traditional dichotomies of subject and object, mind and body, consciousness and the world. An ironic "self" would be a partial or occasional one—as Leo Bersani has observed, "Irony is the style of a mind in constant metamorphosis"—and could not be privileged in the way that skepticism might privilege "consciousness" as a secure point of arrest in the life of the subject. Irony thus emerges as a second trope useful in characterizing the speculatively mobile form of thought that takes the middle voice as its precursor.[16] A practice of rhetorical invention that permanently resists closure would be both metonymic and ironic: metonymic to the extent that the activity of thinking can become a palpably erotic experience whose pleasure does not depend on its affiliation with a traditional satisfaction; and ironic because it nevertheless recognizes that the present satisfaction is insufficiently pleasurable to exhaust desire, which must therefore always intend beyond itself. Taken together, irony

15. See the remarks appended to Paul de Man's essay "Nietzsche's Theory of Rhetoric," *Symposium* 28, no. 1 (1974): 45.

16. See Leo Bersani's discussion of Lautréamont in *A Future for Astyanax: Character and Desire in Literature* (Boston: Little, Brown, 1976), p. 197. For an example of the pertinence of "irony" as a descriptive term, consider the following remarks by Denis Donoghue on Jacques Derrida's *Of Grammatology:* Derrida "does not claim to have stepped beyond metaphysics but rather to have read the metaphysicians in a spirit of suspicion. If we were speaking naively within the philosophic terms, we would say that he is a sceptic, but that term has meaning only within a naive relation between mind and concept. Derrida's spirit is more properly called ironic. Irony smiles upon contradictions and speaks blithely of catastrophe: it dislikes residence and offers itself, like Derrida's work, as a philosophy for nomads." Donoghue continues, "Derrida circumvents residence by resorting to the concept of *jeu* as an act logically prior to the possibility of presence or absence. The final intention of *De la grammatology*, according to Derrida, is 'to make enigmatic what one thinks one understands by the words 'proximity', 'immediacy', and 'presence'. Could any stated aim express the spirit of Irony more precisely?" See Donoghue, *Ferocious Alphabets* (Boston: Little, Brown, 1981), p. 157.

and metonymy suggest an aleatory procession of thoughts able to delight in the immediate situation at the same time that it looks forward to further desiring movement.

Such an activity of continually renewed invention finds early theoretical support in Coleridge's 1817 *Treatise on Method*, a work in which Coleridge proposes to examine diverse forms of discourse in terms of their appropriate use and without assigning to any particular discourse a preeminent ranking. The various literary and philosophic discourses available to a speaker are to be evaluated, Coleridge says, according to "the relations which things bear to each other, or to the observer, or to the state and apprehension of the hearers."[17] This allusion to the logical, ethical, and pathetic dimensions of persuasion suggests a situational, context-specific characterization of utterance, and consequently, a theory of invention based on improvisation and experiment rather than simple ratification of a privileged tradition. No single "method" is privileged, then, over all others. Coleridge can speak of the philosophic method of Plato, the scientific method of Bacon, and the poetic method of Shakespeare without implying a hierarchical ranking among philosophy, science, and poetry. Consistent with the view that no discourse provides more than a temporarily adequate account of reality, Coleridge argues in effect for a proliferation of methods. If the meaning of the world must be renewed at every occasion, then the efficacy of a method is not given once and for all but depends instead on the creative intervention of the speaker.

The contemporary project for a renewal of the middle voice can be affiliated, in this sense, with the Romantic tradition. In her 1926 essay "Composition as Explanation," Gertrude Stein appropriates the term *Romanticism* to describe her own writing. For Stein, Romanticism implies a willingness to remain open to the irreducible character of the present occasion, an affirmation of novelty or difference as a positive virtue: "Romanticism is then when everything being alike [because one resists the temptation to construe the world according to a tradition hierarchy of values] everything is naturally

17. Samuel Coleridge, *Treatise on Method as Published in Encyclopedia Metropolitana*, ed. Alice Snyder (London: Constable, 1934), p. 26.

simply different."[18] By thus affirming the experiential present, the Romantic would bridge that most fundamental of cleavages, the one inaugurating this world (whose attributes are negative ones: fragmentation, temporality, difference) and the next (locus of presence, fullness, and the eternal). But in order to recover "paradise" it is necessary to recognize that presence is constituted in a differential relationship with absence; both presence and absence are constituents of reality. As William Blake observes in *The Marriage of Heaven and Hell,* "without Contraries is no progression. Attraction and Repulsion, Reason and Energy, Love and Hate, are necessary to Human existence." Although there can be a practice of presence, presence can never be an achieved state.[19] Properly understood, "paradise" would embrace both those moments when desire succeeds in obtaining satisfaction and those other times when the recalcitrant suchness of the world breaks in on the subject's complacent self-enclosure.

For this reason, we cannot, as Ezra Pound had hoped to do in *The Cantos,* "write Paradise." That is, we cannot know the meaning of our lives once and for all. To try to do so, to reside exclusively in an artificial paradise of fixed and abiding meanings would only entail sacrificing the one authentic "presence" available to us: the suchness beyond and enclosing the realm of concepts. William Blake will therefore insist that "Energy is the only life, and is from the Body; and Reason is the bound or outward circumference of Energy. Energy is Eternal Delight." Paradise resides in the unending genesis of desire within the body, rather than in any one of the temporary satisfactions encountered by the self in its progress through history. Pound says as much elsewhere in *The Cantos:* "Le Paradis n'est pas artificiel / but is jagged"; beauty must alternate with the sublime in

18. Gertrude Stein, "Composition as Explanation," in *Selected Writings,* p. 520.

19. William Blake, *Complete Poetry and Prose,* ed. G. Keynes (London: Nonesuch Library, 1975), p. 149. In *Civilization and Its Discontents,* Freud says that "what we call happiness in the strictest sense comes from the (preferably sudden) satisfaction of needs which have been dammed up to a high degree, and it is from its nature only possible as an episodic phenomenon. When any situation that is desired by the pleasure principle is prolonged, it only produces a feeling of mild contentment. We are so made that we can derive intense enjoyment only from a contrast and very little from a state of things" (*SE,* 21:76).

an unending process of metamorphosis. Forgoing the sadomasochistic dream of perfect mastery, *homo sapiens* must become, in Huizinga's phrase, *homo ludens,* playing only for temporary advantage. Pound describes this movement of thought as a "periplum, not as land looks on a map / but as sea bord seen by men sailing"[20]; instead of a simultaneity promising definitive knowledge, an endless emergence of meaning from just over the horizon.

In his *Biographia Literaria,* the work immediately preceding the *Treatise on Method,* Coleridge develops a similar characterization of knowledge. For Coleridge, the subject and object of knowledge are interanimating and interdependent. Knowledge should be understood, therefore, as an activity, its closure at most an operative or enabling fiction. The well-known distinction between Imagination and Fancy represents in this sense the difference between genuinely inventing on the spur of the moment and the mere rote repetition of tradition. The concept of occasion is thus crucially important for Coleridge's theory of invention: "On the immediate which dwells in every man, and on the original intuition or absolute affirmation of it . . . all the certainty of our knowledge depends; and this becomes intelligible to no man by the ministry of mere words from without."[21] Knowledge is produced not by a process of gradual approximation to a preexistent truth, but in response to the speaker's present situation, whose endless novelty insures that our experience of the world will never finally be resumed in the nets of language. In other words, knowledge develops as a form of action within the wider sphere of ultimately inexpressible human being.

Coleridge frequently glosses his conception of knowledge as constantly renewed invention with an analogy to organic process. In the following passage, the antithetical pairing of Imagination and Fancy is restated in terms of the organic and the inorganic:

> The form is mechanic when on any given material we impress a predetermined form not necessarily arising out of the properties

20. Quotations of Ezra Pound, *The Cantos* (New York: New Directions, 1972) are from Cantos CXX, XCII, and LIX. Blake, "The Marriage of Heaven and Hell," p. 149.

21. Samuel Coleridge, *Biographia Literaria,* ed. George Watson (London: J. M. Dent and Sons, 1971), p. 140.

> of the material. . . . The organic form, on the other hand, is innate; it shapes as it develops itself from within, and the fullness of its development is one and the same with the perfection of the outward form.[22]

This description of knowledge as generative rather than simply imitative recurs in the *Treatise on Method* when Coleridge defines *method* as "a way, or path, of transit," "a progressive transition from one step in any course to another," and argues that "where the word Method is applied with reference to many such transitions in continuity, it necessarily implies a Principle of UNITY WITH PROGRESSION,"[23] or permanence in the midst of change; a process of adapting past to present that is equivalent to organic growth.

Coleridge's insistence that "method" depends on a principle of unity with progression usefully underscores the fact that a renewed form of the middle voice would unfold as a dynamic of repetition and difference in which the genuinely novel and unprecedented is continually converted into reassuring familiarity. Understood as a principle of repetition, metaphor therefore emerges as a third trope necessary to complete this characterization of the middle voice. But unlike metonymy or irony, which comprise only relative moments within the middle, metaphor would seek constantly to subvert the middle voice by transforming its disclosive potential into the self-enclosed equilibrium of the reflexive.[24] The traditional definition of metonymy—as a movement, for instance, from cause to effect—

22. See *Coleridge on Shakespeare*, ed. Terence Hawkes (Harmondsworth, Middlesex: Penguin, 1969), p. 77.

23. Coleridge, *Treatise on Method*, p. 2.

24. For an example of such a transformation, consider Friedrich Schiller's definition of the "play impulse" in his *Letters on the Aesthetic Education of Man*. According to Schiller, the play impulse can be understood as a harmonious interchange between a spiritual principle of subjective will (the "form impulse") and a bodily principle representing human rootedness in history (the "sense impulse"). He notes that the play impulse "would aim at the extinction of time *in time*, and the reconciliation of becoming with absolute being, of variation with identity. The sense impulse wants to *be* determined, to receive its object, the form impulse wants to determine for itself, to produce its object; so the play impulse will endeavor to receive as it would itself have produced, and to produce as the sense aspires to receive." Schiller thus imagines an ideal condition of perfect correspondence in which the opposition of spirit and matter, Being and Becoming, the self and its environment has tacitly been resolved in favor of the subject of knowledge: when

transpires within the space of an already constituted unity, just as two-term irony—where the speaker intends the opposite of what is manifestly said—implies a closed universe of meanings. Both irony and metonymy presuppose the metaphoric convergence of meanings conventionally apprehended as "reality." The middle voice is thus imperfectly aleatory: as soon as the "present occasion" becomes an object of consciousness, it has already been replaced by an idea, or metaphorically equated with the past. The present that can be *known* is a construct. Inflected from a relative point of view, it is infused with the speaker's desire for permanent equilibrium.

If the middle voice can be understood as an endless alternation of beauty with the sublime, it must nevertheless be recognized, then, that beauty threatens constantly to erase the irreducibly specific character of sublimity. To revert for a moment to the vocabulary of the introductory section of *Kaironomia, kairos* always defers in the end to *nomos;* the unforeseen opportunity of the present occasion is assimilated by the already known. In *The Pleasure of the Text,* Roland Barthes contrasts beauty and sublimity, or "pleasure" and "bliss," in this way:

> Text of pleasure: the text that contents, fills, grants euphoria; the text that comes from culture and does not break with it, is linked to a *comfortable* practice of reading. Text of bliss: the text that imposes a state of loss, the text that discomforts (per-

matter and spirit have finally been reconciled, "Man will combine the greatest fullness of existence with the utmost self-dependence and freedom, and instead of abandoning himself to the world he will rather draw it into himself" so that "the whole infinity of its phenomena" will submit "to the unity of his reason." On the other hand, Schiller realizes that a perfect "equilibrium of reality and form" is an unattainable ideal: "there will always be a preponderance of one element or the other, and the utmost that experience can achieve will consist of an oscillation between the two principles." In a letter to Goethe (16 October 1795), he describes his own compositional practice in the *Letters* precisely in these terms as a restless dynamic: "While the philosopher may allow his imagination and the poet his power of abstraction to rest, I am obliged . . . to maintain both of these powers in an equal state of tension, and only by a constant movement within me can I keep the two heterogeneous elements in a kind of solution"; see *Letters on the Aesthetic Education of Man,* trans. Reg Snell (New Haven: Yale University Press, 1954), pp. 74, 69, 81, 6.

> haps to the point of a certain boredom), unsettles the reader's historical, cultural, psychological assumptions, the consistency of his tastes, values, memories, brings to a crisis his relation with language.

Barthes thus defines "bliss" as a disruptive intrusion into the psychic equilibrium of the self, an intrusion that unsettles a traditional circuit of satisfactions. But the term "bliss" is itself already interpreted from the standpoint of pleasure. Pleasure encroaches on bliss from the beginning:

> *Pleasure/Bliss:* terminologically, there is always a vacillation—I stumble, I err. In any case, there will always be a margin of indecision; the distinction will not be the source of absolute classifications, the paradigm will falter, the meaning will be precarious, revocable, reversible, the discourse incomplete.[25]

The uncertain character of Barthes's distinction between pleasure and bliss helps bring into focus a paradox that is in fact inevitable in any encomium on the middle voice. To put it briefly, the middle voice would dispute the notion of correspondence even as it claims to correspond with "the mutability of Being." In other words, as soon as it recognizes its own mobility, the "middle voice" merges with the reflexive, mistaking a static image for an unpicturable event. The middle must always remain alert, therefore, to the danger of falling prey to the complacent self-delusion of the reflexive voice. For this reason, the middle voice can never be an achieved state. Only as a decision to programmatically confound expectation can it make possible a state of wakefulness resistant to dogmatic slumber. Accordingly, in the remaining Argument to be made in this second section of *Kaironomia,* the "middle voice" will no longer refer to a specific grammatical form or ratio of tropes. The "middle voice" will stand instead for the *will-to-invent.*

25. Roland Barthes, *The Pleasure of the Text,* trans. R. Miller (New York: Hill and Wang, 1975), pp. 14, 4. The vacillation within Barthes's terminology is even more striking in the original French: at once the lack or difference provoking invention, "jouissance" (rendered in English as "bliss") has also as one of its meanings "sexual coming."

II

Modernism in the arts has often been said to represent a desire to invent not on the basis of received knowledge but in response to the irrational novelty of the moment. According to this view, the modernist would even abolish tradition in order truly to be present, forever engaged in the process of conferring meaning on the world. But the aspiration for modernity very soon discovers that speculative invention can never be freed once and for all from the burden of the past. Paul de Man describes it this way in *Blindness and Insight:* "If history is not to become sheer regression or paralysis, it depends on modernity for its duration and renewal; but modernity cannot assert itself without being at once swallowed up and reintegrated into a regressive historical process."[26] The modern artist is thus suspended between mutually exclusive alternatives. On the one hand, there is the desire for ceaseless change; on the other, the recognition that any act of invention is historically determined, part of the very tradition it seeks to transcend.

As de Man has pointed out, Friedrich Nietzsche provides an exemplary instance of the modernist sensibility, so self-divided, finally, that his thinking characteristically emerges in conflict with itself. Nietzsche will first of all challenge the conventional wisdom according to which knowledge is understood as a copy of reality. In this passage from his 1872 essay "On Truth and Lie in an Extra-Moral Sense," he emphasizes the genuinely productive nature of knowing:

> On the whole it seems to me that the "right perception"—which would mean the adequate expression of an object in a subject—is a nonentity full of contradictions for between two utterly different spheres, as between subject and object, there is no causality, no accuracy, no expression, but at the most an *aesthetical* relation, I mean a suggestive metamorphosis, a stammering

26. Paul de Man, *Blindness and Insight* (New York: Oxford University Press, 1971), p. 151. See also Thomas McFarland's discussion of the "originality paradox" in his *Originality and Imagination* (Baltimore: Johns Hopkins University Press, 1985).

> translation into quite a distinct foreign language, for which purpose there is needed at any rate an intermediate sphere, an intermediate force, freely composing and freely inventing.[27]

For Nietzsche, knowledge does not so much imitate a preexistent truth as it invents a new one. The nature of immediacy is therefore fugitive and discontinuous. That is, the experiential present is constantly assimilated to an alien tradition. Nietzsche thus proposes an essential discrepancy between knowledge and reality: "Every idea originates through equating the unequal. . . . The disregarding of the individual and real furnishes us with the idea, as it likewise gives us the form; whereas nature knows of no forms and ideas, and therefore knows no species but only an *X*, to us inaccessible and indefinable."[28] Truth and falsehood should be understood, then, as figures of thought rather than properties of reality. Nietzsche's theory of rhetorical invention represents a revival in this sense of the sophistic view that knowledge derives not from correspondence with the truth but, as Nietzsche says, "from purely rhetorical tricks and devices."[29]

Since knowledge is fundamentally the result of rhetorical artifice, the meaning of the world is always contingent on a given historical situation. Nietzsche therefore counsels an experimental attitude, or

27. Friedrick Nietzsche, *Early Greek Philosophy and Other Essays*, trans. M. A. Mügge, vol. 2 of *The Complete Works of Friedrich Nietzsche*, ed. Dr. Oscar Levy (New York: Macmillan, 1909–11), p. 184.

28. Ibid., pp. 179, 180. In the founding text of modern evolutionary theory, Charles Darwin makes a similar point when he says that he regards the term *species* "as one arbitrarily given for the sake of convenience to a set of individuals closely resembling each other, and that it does not essentially differ from the term variety, which is given to less distinct and fluctuating forms. The term variety, again, in comparison with mere individual differences is also applied arbitrarily, and for mere convenience sake." For Darwin, in a world that is constantly changing, there are no "hard edges" on which knowledge might fasten once and for all; see *The Origin of Species*, ed. J. W. Burrow (Harmondsworth, Middlesex: Penguin, 1968), p. 108.

29. Quoted by de Man in "Nietzsche's Theory of Rhetoric," p. 35. De Man remarks there that Nietzsche's "straightforward affirmation that the paradigmatic structure of language is rhetorical rather than representational or expressive of a referential, proper meaning . . . marks a full reversal of the established priorities which traditionally root the authority of language in its adequation to an extra-linguistic referent or meaning, rather than in the intralinguistic resources of figures."

one that will multiply the occasions of invention. Refusing to privilege tradition, in the *The Gay Science* Nietzsche argues that our beliefs must "descend to the modesty of a hypothesis, of a provisional experimental point of view, of a regulative fiction."[30] For Nietzsche, tradition must always be put to the test. It provides no more than a point of departure for an activity of rhetorical invention that programmatically differs from the past. Nietzsche's decision, as it were, always to be elsewhere would thus ensue in a constant mutation of thought. His theory of rhetorical invention implies a "practice of presence," a phrase that can be understood in either of two ways. First of all, presence is a practice to the extent that the illusory "presence" of meaning is not unproblematically given but must be produced. Second, a practice of presence would consist of a continual return to the occasion of thought, the indifferent specificity of the immediate situation.

But when Nietzsche renews the sophistic view of language in a second important respect, he concedes in effect that a utopian aspiration underlies his program for endless invention. In a tacit reference to the doctrine of the *dissoi logoi,* Nietzsche completes his characterization of knowledge as rhetorical artifice with the following remark: although the antithesis of individual and species "does not come from the essence of things, . . . on the other hand we do not dare to say that it does not correspond to it; for that would be a dogmatic assertion and as such just as undemonstrable as its contrary."[31] The paradox to which Nietzsche alludes might be stated this way: to the proposition that existence is nonsensical prior to the imposition of a human fabric of meanings, must be added the qualification that "nonsense" takes its meaning from a contrast with "sense"; so that for life to be meaningless, it must also be not nonsensical. In other words, "this" is meaningful nonsense. Nietzsche's argument thus implicitly affirms the identity of opposites. Concerning the scientific passion for experimentally derived "truth," Nietzsche says that "it is still a *metaphysical* faith upon

30. Nietzsche, *The Gay Science,* trans. Walter Kaufmann (New York: Random House, 1974), p. 280.

31. Nietzsche, *Early Greek Philosophy and Other Essays,* p. 180.

which our faith in science rests."[32] Even the intention to be intentionless represents a dogmatic choice. Every new idea is affiliated with a tradition of meanings, a network of diacritical oppositions that preclude the possibility of genuine originality. "Skepticism" and "dogmatism" mutually embrace one another: conceptual metamorphosis depends on inflexible self-repetition, just as dogmatic certainty is already undermined by the prospect of doubt. As William James observes in *The Principles of Psychology,* "Into the awareness of the thunder [that breaks upon consciousness] the awareness of the previous silence creeps and continues; for what we hear when the thunder crashes is not thunder *pure,* but thunder-breaking-upon-silence-and-contrasting-with-it."[33]

The doctrine of the *dissoi logoi,* implicitly invoked in the course of Nietzsche's argument on behalf of interpretive mobility, figures even more prominently in the writings of Jacques Derrida. In a line of thought strikingly reminiscent of Gorgias's treatise "On Not-Being," Derrida has argued that any "metaphysics of presence," or belief in a realm of immutable Truth, is dependent on an unacknowledged "outside" in relation to which the "inside" constitutes itself. The "outside," whose existence is denied by the "inside's" claim of absolute coincidence with Reality, is precisely the guarantor of the identity of the "inside." Hence the following paradoxical chapter title from *Of Grammatology:* "The Outside X the Inside." Being and Not-Being, Reality and Illusion, Truth and Lie, mutually presuppose one another. Thus, a metaphysics of presence, which *spatializes* reality in the form of a knowable simultaneous plenitude, depends for its very existence on that which it would negate altogether, namely, *temporality,* or the endlessly differentiating flux of history. As Saussure recognized in his *Course in General Linguistics,* there are no positive terms in language, only differences. The meaning of a word is defined not positively by its content but negatively in relation to what it is not. Derrida's term for the relational nature of meaning is *différance.* For Derrida, *différance* constitutes language as precisely a play of differences. It defines the beginning

32. Nietzsche, *The Gay Science,* p. 283.
33. James, 1:240–41.

possibility of meaning as the difference between form and content, word and idea, consciousness and the world. *Différance* opens up a space within or through which "mind" apprehends "reality." Our knowledge of the world thus depends on a network of contrasting terms not given in nature. Language inflects the world according to a logic of diacritical oppositions arbitrarily imposed on natural fact.[34]

But there is a further and more crucial consequence to be extrapolated from Saussure's insight concerning the relational nature of meaning. In addition to differentiating terms, *différance* also introduces a temporal delay or deferral of meaning. That is, mutually presupposed opposites never coincide in time. Contrary possibilities can never be truly contemporary. According to Derrida, the present fullness of meaning is therefore merely a relative moment. The meaning we attribute to the world is always a historically contingent one emerging against a presupposed backdrop of alternative meanings that is not itself available as an object of knowledge. Our present understanding of the world is thus already destined to be superseded. No single interpretive standpoint can be privileged. The present occasion of rhetorical invention inevitably comprises a limited point of view. In fact, every attribution of meaning is unforeseeably one-sided. Diacritically related to a network of absent meanings, the act of invention can never grasp the extent to which it is a product of history.

The possibility of an objective appraisal of the immediate situation is precluded, then, not simply by the mutability of Being but because of the diacritic/diachronic character of understanding. Within language the individual meaning, viewed paradigmatically, is produced in a differential relationship with an absent meaning, and viewed syntagmatically, is never "present" to consciousness but has always already just slipped away. The antithetical nature of truth thus insures that the object of knowledge can never be disclosed once and for all. Derrida would therefore concur with Nietzsche on the provisional character of the *logos* of the moment: "Against positivism, which halts at phenomena—'There are only

34. Jacques Derrida, *Of Grammatology*, trans. Gayatri Chakravorty Spivak (Baltimore: Johns Hopkins University Press, 1976). See also Ferdinand de Saussure, *Course in General Linguistics*, trans. Roy Harris (London: Duckworth, 1983).

facts'—I would say: No, facts is precisely what there is not, only interpretations."[35] For Derrida, differentiation proliferates without limit. Knowledge unfolds as a dialectical movement of thought whose mutation over time cannot be reduced to a fundamental structuring opposition. "Reality" is, finally, an endless slippage of meaning that conforms to no predictable itinerary.

But the characteristic strategy of thought is to live as if the relative were the absolute, as if knowledge could secure a purchase on unconditioned truth. Derrida suggests that this longing for a centered universe of meanings represents a willful forgetting of the reality of *différance*.[36] Humanity habitually overlooks the productive character of knowing in order to reside within a world of reassuring certainties. For Derrida, cultural tradition has become a perpetual hindrance to creative interaction with the world. Language itself continually obscures the specific character of the present. From this point of view, the task facing modern consciousness therefore would be a "resurrection of the body," a reaffirmation of the ambivalent and protein vitality that constitutes the foundation of human being. To recognize language as essentially contradictory or "nonsensical" would enable our interpretation of the world to transpire as a form of erotic play, as one aspect of an ongoing process that must in principle elude definitive verbal description. No longer externalized in reality,

35. Friedrich Nietzsche, *The Will to Power*, trans. Walter Kaufmann and R. J. Hollingdale (New York: Random House, 1967), p. 267.

36. According to Derrida, the mind attempts to evade the unpleasant reality of *différance* by interpreting the psychophysical experience of speech as confirmation of the self's presence to itself. This privileging of speech as token of timeless, unmediated meaning entails a corresponding devaluation of the written word which is denied any specific attributes as a medium of communication: writing is merely an instrument for the representation of speech. That is, in order to entertain the fantasy of enduring truth, the mind must obscure the revelation writing affords of thought as a historical process. As Jack Goody and Ian Watt argue in their essay "The Consequences of Literacy," the invention of writing undermines the possibility of apprehending "the cultural tradition as any sort of patterned whole"; see *Comparative Studies in Society and History* 5 (1963): 335. What Derrida calls "logocentrism" thus looks back to an idealized preliterate condition. In this sense, his "grammatological" analysis of the history of consciousness as the repression of writing can be compared to that persistent nostalgia for tribal holism Karl Popper indicts in *The Open Society and Its Enemies* (Princeton: Princeton University Press, 1950).

our various forms of discourse could then be grasped as evolving instruments of use, or perhaps, as an array of equally valid, equally likely structures of experience. As Gayatri Chakravorty Spivak comments, "we must learn to use and erase our language at the same time."[37] Instead of the *tragedian,* who "lives like an exile the necessity of interpretation," Derrida advocates the way of the *farceur,* which would entail a skeptical withholding of assent, an ironic detachment from the meaning attributed to reality, an affirmation "of the freeplay of the world . . . without truth, without origin, offered to an active interpretation."[38] Thus initiating "a strategy without finality," the farceur would continually invent the meaning of the world anew.[39]

This suspicion of received knowledge does not mean, however, that rhetorical invention can remain free from entanglement with tradition. For Derrida, the desire for a centering presence is a constitutive feature of language itself. Meaning necessarily depends on a hierarchical ordering of paired terms. Derrida cannot therefore unequivocally endorse the intention always to be engaged in an activity of renewing the meaning of the world. According to the logic of *différance,* even the recognition of the relativity of every standpoint comprises a relative point of view. No state of consciousness can know the extent to which it is constituted by an unguessed-at other. There is thus no escape from metaphysics. In an inevitable gesture, Derrida qualifies his argument on the antithetical nature of truth with this remark: "For us, [*différance*] remains in metaphysical name; and all the names that it receives from our language, so far as they are names, are metaphysical. . . . What is unnameable is the play that brings about the nominal effects, the relatively unitary or atomic structures we call names."[40] Since every assertion represents the imposition of a partial truth, Derrida must modify his own

37. Gayatri Chakravorty Spivak, "Translator's Preface" to Jacques Derrida's *Of Grammatology,* p. xviii.

38. Jacques Derrida, "Structure, Sign and Play in the Discourse of the Human Sciences," in Macksey and Donato, eds., *The Structuralist Controversy,* p. 264.

39. Jacques Derrida, "Differance," in his *Speech and Phenomena,* trans. David Allison (Evanston, Ill.: Northwestern University Press, 1973), p. 135.

40. Ibid., pp. 158, 159.

aspiration for endless invention. Recognizing the mediated character of perception does not exclude the possibility that knowledge truly corresponds with the world. As Spivak says, "The playfully disseminating rather than proprietorially hermeneutic gesture of interpretation (n)ever penetrates the hymen of the text."[41] In this connection, Derrida compares *différance* to the middle voice, which "precedes and sets up the opposition between activity and passivity," the opposition, in other words, between genuine originality and received tradition.[42]

The following passage is Chuang-tze's version of the paradox that threatens to empty Derrida's argument of its positive content.

> There was a beginning. There was a time before that beginning. And there was a time before the time which was before that beginning. There was being. There was non-being. There was a time before that non-being. And there was a time before the time that was before that non-being. Suddenly there is being and there is non-being, but I don't know which of being and non-being is really being or really non-being. I have just said something, but I don't know if what I have said really says something or says nothing.[43]

The diacritical oscillation of contraries thus extends to the very statement that would express the contradictory nature of reality. Even the position that claims to be no-position is haunted by nostalgia for definitive truth. *Différance* becomes an equivalent term for what Edward Said has referred to as the pure intransitive beginning of consciousness, "perhaps our permanent concession as finite

41. Spivak, "Translator's Preface," p. lxvi.

42. Derrida, "Differance," p. 130. Derrida pursues the analogy between *différance* and the middle voice later in the same essay when he says that *différance* "announces or rather recalls something like the middle voice, . . . it speaks of an operation which is not an operation, which cannot be thought of either as a passion or as an action of a subject upon an object, as starting from an agent or from a patient, or on the basis of or in view of, any of these *terms*. But philosophy has perhaps commenced by distributing the middle voice, expressing a certain intransitiveness, into the active and passive voice, and has itself been constituted in this repression" (p. 137).

43. Chuang-tze, in Chan, p. 185.

minds to an ungraspable absolute."[44] Derrida's intention must therefore remain a utopian one. The farceur's freedom from nostalgia and self-serving delusion is a perpetual task. Even the recognition of the differential nature of meaning issues from a tragically divided consciousness.

The knowledge that farce provides is thus tragic in spite of itself. The strategy of the farceur is already the strategy of the tragedian. Even the farceur who would avoid error by saying nothing other than "the contradictory nature of reality" still affirms a relative point of view. If one cannot invent the meaning of the world once and for all, neither can one remain free from attachment to a partial truth. The euphoric prospect of endless interpretation in fact depends on continuous involvement in relative historical situations. A speculatively mobile form of invention would thus delude itself even as it remains skeptically alert. In other words, the activity of rhetorical invention will conform to a contradictory injunction. At the same time that it programmatically denies the possibility of definitive truth, it will nevertheless dream of once and for all deciphering the meaning of the world.

Michel Foucault's *The Order of Things* can be understood as an example of an argument caught in just such a double bind. In *The Order of Things,* Foucault recounts the slow dissolution of belief in a closed universe of fixed and abiding meanings, the gradual recognition of the historical character of understanding. Foucault divides European cultural history into three periods, roughly corresponding to the fifteenth and sixteenth centuries (the Renaissance); the seventeenth and eighteenth centuries (the Classical age); and the nineteenth and twentieth centuries (the modern era). Each period is then described in terms of a characteristic set of underlying epistemological assumptions. According to Foucault, in each period these assumptions have been inflected by a desire to release consciousness from the burden of invention, a desire for a permanent state of equilibrium that ensues in a persistent repression of the reality of persuasion. In order to be consistent with his own desire for endless

44. Edward Said, *Beginnings: Intention and Method* (New York: Basic Books, 1975), p. 77.

interpretation, Foucault is therefore compelled to underline the provisional, purely rhetorical nature of the argument he himself makes in *The Order of Things*. Thus, even as he argues for the inevitability of speculative thought, Foucault will display the historically relative character of this point of view.

In his exploration of the assumptions governing pre-seventeenth-century thought, Foucault discovers that the mind of the Renaissance is able to render words and things an indivisible unity by attributing a tropical design to the material world. During the Renaissance, metaphor acquires, as it were, an ontological value. The similitudes of language are one with the similitudes of reality. The principle controlling the elaboration of knowledge in this period is, then, *resemblance*. Resemblance (formal likeness, similarity, or correspondence) provided the means for Renaissance Europe to knit the fabric of reality. For the age that preceded the Classical deployment of a picture theory of language and the modern invention of historical modes of explanation, knowledge was founded on the proposition that to look for the meaning of something "is to bring to light a resemblance."[45] A thing exists only insofar as it resembles another thing, which in turn resembles a third, and so on in a cycle of endless recursions as the world folds back upon itself in layer upon layer of numberless likenesses. Foucault describes this world of endless resemblances as "plethoric": the universe is like itself in an infinite number of ways. There is always a hidden resemblance waiting to be discovered. In this sense, the possible meanings of reality can be multiplied without limit. But the discovery of resemblance promises no truly improvisational or occasional form of invention. Resemblance implies repetition. For the mind of the Renaissance, knowledge depends on conformity with a privileged origin of meaning. Reality is therefore organized into a hierarchy of similitudes, a Great Chain of Being whose elements to varying degrees approximate to a fixed point of return. The positing of an authorizing center of meaning thus converts invention into simple amplification. Instead of a succession of encounters with genuine novelty, knowledge unfolds as an endless return of the same.

45. Michel Foucault, *The Order of Things* (London: Tavistock, 1970), p. 29.

According to Foucault, at the beginning of the seventeenth century a mutation occurs that simultaneously disrupts and reaffirms the Renaissance belief in the coincidence of Being and Knowledge. In his investigation of the Classical experience of language, Foucault discovers a theory of signification which posits a "pure and simple connection between what signifies and what is signified." Although the Classical age separates consciousness from the world, it subordinates one to the other in a representational closure: "In fact, the signifying element has no content, no function, and no determination other than what it represents: it is entirely ordered upon and transparent to it." The Classical age thus converts resemblance into a systematic effect obtaining between the order of Being and the order of Knowledge. Knowledge is understood as mirroring correspondence, or a transparent window on nature. Since the signifier is transparent with respect to the signified, the relationship between consciousness and the world is not dialogic but one of direct transcription. In the ideal utterance, natural fact would be rendered with an immediacy approaching that of sense experience: "The table of the signs will be the *image* of the things." The Classical theory of language can in fact be said to derive from a model of sense experience. As the saying goes, we *behold* the truth. Language ideally approximates in its self-effacing transparency to the unmediated apprehension we like to think characteristic of the life of the senses.[46] The Classical order of things is therefore necessarily unaware of itself as a form of persuasion. Truth depends on correspondence with a reality outside language rather than on the sensuous properties of the medium itself. Foucault observes that this is "the age in which the poetic dimension of language is defined by metaphor, simile, and allegory."[47] That is, in order to enjoy the illusion of unmediated meaning, the Classicist must establish a hierarchy of forms of discourse

46. Ibid., pp. 67, 64, 66. In *The Nouveau Roman: A Study in the Practice of Writing* (London: Elek, 1972), Stephen Heath suggests the representational conventions of the realistic novel have their origin in the Classical theory of language: Balzac's novelistic practice in *The Human Comedy* is based on the assumption that words are "self-effacing in the process of the presentation of things." For Balzac, the language of fiction "is the quicksilver of a mirror that directly reflects in image the tableau of Society" (p. 18).

47. Foucault, *The Order of Things*, p. 51.

that privileges the plain style over other deployments of language. The theory of the sign as transparent in the process of representation thus precludes a rhetorical understanding of the multiplicity of persuasive strategies.

Where the Renaissance conceived of Being and Knowledge as an indivisible unity, and the Classical age posited a representational duality, the discovery of history at the beginning of the modern era forever dispels the possibility of a lasting communion between consciousness and the world. The Classical age was able to represent Being as a spatial simultaneity only so long as it ignored temporal process. The discovery of history provokes the recognition that knowledge depends on the context in which it is produced. As Heidegger would argue in *Being and Time,* the world is properly understood as a nonobjectifiable, all-encompassing horizon within which humanity is situated. The attribution of meaning is always contingent, therefore, on a particular set of circumstances. Knowledge unfolds as a succession of discrete occasions of invention that are not arranged according to any teleological scheme. In this sense, the coincidence of Being and Knowledge is forever deferred. Being is dissolved, as it were, in a complex dispersion of forms of discourse. Foucault remarks that language now appears in "a multiplicity of modes," a proliferating array of persuasive strategies emerging in response to the always unforseen opportunity of the immediate situation.[48] The modern era opens onto the prospect of a continually renewed activity of rhetorical invention.

But the will-to-invent immediately encounters an attempt to foreclose the possibility of endless interpretation. Thus, Heidegger's teacher Edmund Husserl believed that in spite of the historical character of understanding, reflective consciousness can transcend the present relative moment and so gain a vantage from which to regard a timeless order of intelligibility. For Husserl, the subject is only apparently enclosed in historically contingent situations. Consciousness can rise above historical circumstance in order to discover a hidden body of assumptions according to which experience is interpreted. Within consciousness there is a realm of transtem-

48. Ibid., p. 304.

poral meanings that will establish knowledge on a certain foundation. Yet at the same time that he affirms this realm of permanent truth outside history, Husserl remains committed to a perspectival, situational characterization of consciousness. In *The Crisis of European Sciences,* he concedes the priority of history when he says that the world "is pregiven to us, the waking, always somehow practically interested subjects, not occasionally but always and necessarily as the universal field of all actual and possible praxis, as horizon."[49] According to this point of view, the meaning of the world depends on an interpretive standpoint that depends in turn on indeterminately further standpoints. The evolving body of assumptions in terms of which the interpreting subject makes sense of its experience cannot in principle become an object of knowledge. Consciousness can never grasp the extent to which the present moment is a product of history. The occasion of invention itself is already an interpretation.

On the one hand, then, Husserl proposes a situational characterization of consciousness according to which thought is inextricably implicated in the historical unfolding of reality. As Merleau-Ponty observes, since "we are in the world, since indeed our reflections are carried out in the temporal flux on which we are trying to seize, there is no thought which embraces all our thought."[50] The thinking animal that repeatedly confuses its identity with historically relative definitions of "human nature" has finally no determinable essence: consciousness can never seize hold of itself from the beginning. On the other hand, Husserl theorizes a situationless subjectivity within which, as he puts it in *Cartesian Meditations,* "the *universal logos of all conceivable being*" will be discovered.[51] Husserl thus attempts to fuse two incompatible notions of history. According to one, time is truly *disclosive,* or productive of genuine

49. Edmund Husserl, *The Crisis of European Sciences and Transcendental Phenomenology,* trans. David Carr (Evanston, Ill.: Northwestern University Press, 1970), p. 142.

50. Maurice Merleau-Ponty, "What Is Phenomenology?" in Joseph J. Kockelmans, ed., *Phenomenology: The Philosophy of Edmund Husserl and Its Interpretation* (Garden City, N.Y.: Anchor, 1967), p. 365.

51. Edmund Husserl, *Cartesian Meditations,* trans. Dorion Cairns (The Hague: Martinus Nijhoff, 1977), p. 155.

novelty. For the other, time is merely *suspenseful,* since its end is never really in doubt. Husserl would in effect reconcile the idea that the world is in consciousness with its contrary, that consciousness is in the world.

This is how Foucault puts the dilemma that Husserl could not overcome:

> The more History attempts to transcend its own rootedness in historicity, and the greater the efforts it makes to attain, beyond the historical relativity of its origin and its choices, the sphere of universality, the more clearly it bears the marks of its historical birth, and the more evidently there appears through it the history of which it is a part . . . ; inversely, the more it accepts its relativity, and the more deeply it sinks into the movement it shares with what is is recounting, then the more it tends to the slenderness of the narrative, and all the positive content it obtained for itself through the human sciences is dissipated.[52]

Being and Knowledge thus exist in an inverse relationship. Knowledge emerges precisely from overlooking the historical specificity of Being. In this sense, the will-to-invent, to make sense of one's experience, depends on a deliberate refusal to acknowledge the irreducible nature of the present. Even an ideally mobile form of interpretation deludes itself as soon as it attributes meaning to the world. The modern aspiration for interpretive mobility is nostalgic in spite of itself for the representational certainty of the Classical age, just as the Classical dualism of Being and Knowledge itself longs for the unity of words and things that prevailed during the Renaissance.

As a historical narrative seeking ultimately to persuade on the basis of its adequacy to the facts, *The Order of Things* cannot evade the desire for definitive truth. Foucault's argument for endless interpretation can itself be understood as an example of the inherently compromised character of the will-to-invent. *The Order of Things* necessarily dissembles itself as experientially given truth. But Foucault can at least call attention to this inevitable deception by underlining the rhetorical artifice of his argument. In other words, do the three stages in Foucault's account of the evolution of Euro-

52. Foucault, *The Order of Things,* p. 371.

pean consciousness represent a genuine temporal progression or should they be understood as an arbitrary imposition on the flux of history? The purely formal nature of the three stages can be expressed in this sequence. First of all, there is the Renaissance confusion of words and things: the mind, as it were, naively perceives "the world." Next, the representational duality of the Classical age: the mind perceives "perception of the world." Finally, the Modern recognition of the relativity of every standpoint: the mind perceives "perception of perception of the world." Marcel Duchamp remarks: "For me the number three is important, but simply from the numerical, not the esoteric point of view: one is unity, two is double, duality, and three is the rest. When you've come to three, you have three million—it's the same thing as three."[53] The discovery of historical process thus results in an infinite regress of interpretations. No longer correspondence with a preexistent truth, knowledge therefore can be understood as a form of persuasion.

In their analysis of Foucault's career subsequent to *The Order of Things,* Hubert Dreyfus and Paul Rabinow suggest that Foucault's "genealogical" approach to the past in, for instance, the multi-volume *History of Sexuality* reflects Foucault's awareness of the contingent character of every historical "truth," his recognition that knowledge is moved by pragmatic concerns. "Genealogy," a term Foucault borrows from Nietzsche, implies the interpreter's involvement in the activity of making sense. Foucault's "history of the present" does not claim to be an objective or disinterested investigation but issues from personally felt urgencies. For Foucault, the "correspondence theory of reality is dead." He cannot, therefore, claim "to give us a true history of the past in the sense of one that is fully adequate to the past, which represents it correctly, which gets the whole picture." The history of the present explicitly and self-reflectively begins with a diagnosis of the current situation." Foucault's genealogical account of the emergence of current cultural practices is meant to enable him not only to understand but, more importantly, to revise the world. Dreyfus and Rabinow thus

53. Pierre Cabanne, *Dialogues with Marcel Duchamp,* trans. Ron Padgett (New York: Viking Press, 1971), p. 47.

allude to a remark by Gilles Deleuze that Foucault "should be seen not as a historian but as a new kind of map-maker—maps made for use not to mirror the terrain."[54]

Foucault's historical reflections in the latter portion of his career were guided by his conviction that pervasive techniques of normalization constitute the greatest threat to human freedom on the contemporary cultural landscape. The past several centuries have witnessed, Foucault says, "an explosion of numerous and diverse techniques for achieving the subjugation of bodies and the control of populations." These techniques define, manipulate, and constrain the possibilities of our lives by compelling us to conform to the deep truths of an illusory "human nature." The challenge confronting us is, therefore, to contrive "a different economy of bodies and pleasures."[55] As Dreyfus and Rabinow maintain, we need to devise cultural paradigms that will "function in such a way as to focus important concerns . . . without preordaining, in a normalizing way, what responses would count as appropriate."[56] To this end, Foucault began an exploration of Classical eudaemonistic philosophers whose "aesthetics of existence," while not strictly replicable in the contemporary world, might nevertheless inspire an alternative "technology of the self"; precisely, an ideally improvisational personal style free from subjection to an arbitrarily restricted vision of what is normally or essentially human.

The History of Sexuality does not, then, claim detached objectivity. It moves from a diagnosis of what Foucault felt is our most pressing current problem to speculation concerning a possible solution. Foucault insists, according to Dreyfus and Rabinow,

> on a pragmatic intent in all significant historiography. . . . That is to say that while the analysis of our present practices and their historical development is a disciplined, concrete demonstration which could serve as the basis of a research program, the diag-

54. Hubert L. Dreyfus and Paul Rabinow, *Michel Foucault: Beyond Structuralism and Hermeneutics*, 2d ed. (Chicago: University of Chicago Press, 1983), pp. 120, 119, 128.

55. Michel Foucault, *The History of Sexuality*, vol. I: *An Introduction*, trans. R. Hurley (New York: Random House, 1978), pp. 140, 159.

56. Dreyfus and Rabinow, p. 198.

> nosis that the increasing organization of everything is the central issue of our time is not in any way empirically demonstrable, but rather emerges as an interpretation. This interpretation grows out of pragmatic concerns and has pragmatic intent, and for that very reason can be contested by other interpretations growing out of other concerns.[57]

Foucault's genealogical assessment of a normalizing society cannot demand assent as an unbiased account. His openly polemical intention is closer to what Richard Rorty has termed "edification," which seeks to "take us out of our old selves . . . to aid us in becoming new beings."[58] For Foucault, every interpretation of the meaning of reality involves the affirmation of ethical and aesthetic values. The *pathos* of the interpreter's situation affects his or her analysis. Foucault thus avows a rhetorical conception of knowledge that envisions a role for other than strictly logical forms of proof. Truth issues not merely from disinterested investigation but involves epideictic praise and blame, as Foucault himself implies in this remark—

> I am fully aware that I have never written anything other than fictions. For all that, I would not want to say that they were outside the truth. It seems plausible to me to make fictions work within truth, to introduce truth-effects within a fictional discourse, and in some way to make discourse arouse, "fabricate," something which does not yet exist, thus to fiction something. One "fictions" history starting from a political reality that renders it true, one "fictions" a politics that does not yet exist starting from a historical truth.[59]

If knowledge is always motivated by the present project of the interpreter, our reception of that knowledge should be no less interested.

57. Ibid., p. xxvi.

58. Richard Rorty, *Philosophy and the Mirror of Nature* (Princeton: Princeton University Press, 1979), p. 360. See also Jean-François Lyotard's discussion of "paralogical" forms of discourse in his *The Postmodern Condition: A Report on Knowledge*, trans. Geoff Bennington and Brian Massumi (Minneapolis: University of Minnesota Press, 1984).

59. Foucault, quoted in Dreyfus and Rabinow, p. 204.

Rather than asking whether or not a particular interpretation corresponds to the facts, the more appropriate question would be: does it provide a fertile *topos* or point of departure for further acts of rhetorical invention?

III

According to a reading left largely implicit in Roland Barthes's *S/Z*, Balzac's short story "Sarrasine" seeks to expose the hidden basis of the desire to confer a lasting meaning on reality. The motive for the attempt to freeze the meaning of the world is revealed in the nature of the relationship between the lover Sarrasine and Zambinella, the beloved who turns out to be a castrato. By establishing an inverse relationship between Zambinella and himself, so that their respective attributes perfectly complement one another, Sarrasine is able to define or fix his own being. Consequently, when Zambinella is later discovered to inhabit an anomalous zone midway between the binary opposition "male/female," Sarrasine is deprived in turn of his own identity. The fixing of the object has, then, as its purpose the desire to fix the subject. Love for the other returns in the form of self-love, or narcissism.[60]

By extension, the attribution of a final meaning to a text must also be considered a narcissistic projection, yet another attempt to release the self from the burden of inventing an original response to the world on the spur of the moment. Seeking to discredit the notion of a canonical reading of a text, Barthes therefore proposes in *S/Z* to posit no definitive interpretation of Balzac's story. In fact, the five "codes" or interpretive *topoi* that define the semantic field of "Sarrasine"—codes of plot, suspense, setting, character, and thematic

60. See Barbara Johnson's discussion of *S/Z* in *The Critical Difference: Essays in the Contemporary Rhetoric of Reading* (Baltimore: Johns Hopkins University Press, 1980). Johnson remarks there that Sarrasine narcissistically regards Zambinella "as his own symmetrical opposite and through her loves only himself. This is why the revelation of the truth is fatal. The castrato is simultaneously outside the difference between the sexes as well as representing the literalization of its illusory symmetry. He subverts the desire for symmetrical, binary difference by *fulfilling* it" (p. 10).

significance—are left completely unstructured with respect to one another. Barthes purposely avoids arranging his commonplace according to a hierarchy that would enable a conclusive reading of "Sarrasine"; "if we make no effort to structure . . . the five codes among themselves, we do so deliberately in order to assume the multivalence of the text, its partial reversibility. We are in fact concerned not to manifest a structure but to produce a structuration."[61] Barthes makes a similar point in his later piece on Edgar Allan Poe's "The Facts in the Case of M. Valdemar":

> There is no univocal determination of the enunciation: in a statement, many codes, many voices, *are there*, without any one dominant. Writing is precisely this loss of the origin, of the source, to the profit of a volume of indeterminacies or of overdeterminations: this volume is precisely that of *signifiance* [a moving play of signifiers that programmatically rejects the possibility of final truth]. Writing arrives exactly at the moment when speech ceases, that is, from the instant when one can no longer ascertain *who is speaking* and when one can only state that *it* [or *Id*] *begins to speak.*[62]

Rather than attributing a magical potency to tradition, consciousness must recognize that every tradition, every body of interpretive *topoi*, represents a historically relative apprehension of the meaning of reality. Barthes would resist the self-serving character of interpretation by remaining always in motion, hovering above a field of interpretive possibilities without ever coming to rest upon a final meaning.

But can such an intention be any more than a utopian gesture? Can one really take *S/Z* at face value as an ideally disseminated interpretive practice that has transcended once and for all the narcissistic quest after a privileged meaning? As has already been suggested, Barthes's interest in Balzac's story was determined by an

61. Roland Barthes, *S/Z*, trans. Richard Miller (New York: Hill and Wang, 1974), p. 20.

62. Roland Barthes, "Analyse textuelle d'un conte d'Edgar Poe," in Claude Chabrol, ed., *Sémiotique narrative et textuelle* (Paris: Larousse, 1973), pp. 53–54 (my translation).

interpretation of its thematic significance according to which "Sarrasine" addresses a particular interpretive dilemma, namely, that the meaning of the world is more the product of the observer's self-serving rhetorical artifice than a property of reality itself. In the interpretation of "Sarrasine" presupposed by Barthes, the relationship between Zambinella and Sarrasine is metaphorically equated with the relationship between a text and its interpreter. The thesis to be extrapolated from "Sarrasine" could therefore be phrased this way: the desire to finalize the meaning of a text is equivalent to the desire to master one's self as a controllable object by mastering the Other as a subservient complement.

The fact that such an interpretation underlies *S/Z* has serious consequences for the project in which Barthes claims to be engaged. His assertion that the codes are not hierarchically interrelated is contradicted by his initial subordination of plot, suspense, character, and setting to the thematic significance of Balzac's story.[63] Barthes's interpretive practice, no less than that of Sarrasine, conforms to the proposition that the mind must first project its own order onto the world before it can discover a realm of objects that will satisfy its desire. To borrow a phrase from Frank Kermode, Barthes still craves "the center that will allow the senses to rest."[64] He acquiesces in spite of himself to an arbitrarily privileged meaning. At the moment of his potentially revelatory encounter with the unforeseen spontaneity of Balzac's "Sarrasine," he cannot resist interpreting Balzac's story as a repetition of personal tradition, of his habitual desire not to be determined by habit. The ideal interpretive procedure of *S/Z* thus depends on a crucial element of dissembling; one might say that *S/Z* is a narcissistic gesture pretending to explode narcissism.

63. In his *Structuralist Poetics* (Ithaca: Cornell University Press, 1975), Jonathan Culler notes that "the goals towards which one moves in synthesizing a plot are, of course, notions of thematic structures" (p. 221). Culler says that he is even tempted to affirm that the plot of a work is only the temporal projection of its thematic significance. To the extent, then, that plot is translatable into theme, the meaning of a text can be said to depend precisely on a hierarchical ordering of the interpretive codes.

64. Frank Kermode, *The Genesis of Secrecy: On the Interpretation of Narrative* (Cambridge, Mass.: Harvard University Press, 1979), p. 72.

Even Barthes's ideally speculative practice of rhetorical invention can be understood, then, as an arbitrary imposition of form upon the teeming variety of the world. Knowledge is always interested, or inflected from a relative point of view. There is no escape from interpretive self-delusion. If interpretation depends on a metaphorical equivalence between the attributes of a text and the observer's own interpretive categories, this equivalence always involves a metonymic slippage of meaning. The relationship between text and commentary can never be one of genuine correspondence. Interpretation is a repetition or amplification *in other words* that repeat but do not correspond identically with their object: every interpretation makes a difference. The interpreter, as it were, seduces or leads astray the forever-unavailable proper meaning of the text. Inflected by the values and assumptions of the particular speaker, the act of rhetorical invention is always a self-serving contrivance, the privileging of an idiosyncratic interpretive standpoint. From this perspective, truly mutual understanding, or communication based on a body of premises held in common, becomes a receding possibility. As Seymour Chatman says, "the real world of interpretation is 'partitioned off' . . . ; each of us sees things under different signs, in terms of different categories."[65] Reality is a shimmering mirage, transient and in constant motion, differently perceived by every interpreter.

But if the meaning of a text fundamentally depends on the interpretive strategies brought to bear upon it, the interpreter's history of previous encounters with cultural tradition defines the field of potentially attributable meanings. The interpreted object can be said to control the trajectory of rhetorical invention: even as the mind seeks to impose order upon the world, it is transformed by its objects when it assimilates them to its own developing project. According to Hans-Georg Gadamer, the activity of invention can therefore be compared to the events that transpire on a playing field: "what is intended is the to-and-fro movement which is not tied to any goal which would bring it to an end." The speaker's desire for self-

65. See Seymour Chatman's essay "The Rhetoric of Difficult Fiction," *Poetics Today,* 1, no. 4 (Summer 1980): 26–27.

repetition is only a relative moment within an interpretive movement that continually undermines the attempt to assign a final meaning to reality. In this ceaseless interchange between self and other, it is "the game itself that plays, in that it draws the players into itself and thus becomes the actual subjectum of the playing." Gadamer observes that the original sense of the verb "to play" is "the medial one. Thus we say that something is 'playing' somewhere or at sometime, that something is going on."[66] In other words, the self's interpretive dialogue with its surroundings can be understood in terms of the middle voice as a mutual transformation of subject and object, an endless dynamic of repetition and difference. The ideal activity of rhetorical invention would thus continually renew the meaning of the world in an endlessly different translation of difference into sameness.

Successful adaptation to the mutability of Being depends precisely, Geoffrey Hartman suggests, on an interminable process of "incorporating what continues to violate one's identity."[67] The progress of knowledge can therefore be characterized as a sort of hermeneutical hopscotch in which the interpreter's own tradition of meanings is repeatedly juxtaposed with alien traditions. Even as it gestures toward a timeless space of enduring truth, speculative thought would constantly reopen its investigation of the world, dramatizing the historical character of understanding in an activity of invention that continually disrupts its own attempts at self-enclosure. If all knowledge is unforeseeably one-sided, then acceding to the desire for a permanent state of equilibrium would actually preclude the self's successful adaptation to its environment. Forewarned, then, of the deluded nature of the desire for final truth, the mind must seek constantly to displace itself from its customary centers of meaning. Instead of seeking permanent rest, it must resign itself to a never-ending activity of discharging the desiring tension provoked in the organism in the course of its interaction with its

66. Hans-Georg Gadamer, *Truth and Method*, trans. G. Banden and J. Cumming (New York: Seabury Press, 1975), pp. 93, 446.

67. Geoffrey Hartman, *Criticism in the Wilderness: The Study of Literature Today* (New Haven: Yale University Press, 1980), p. 199.

environment. Knowledge must be apprehensible, as Edward Said says, "in terms of *nomadic centers,* provisional structures always straying from one set of information to another."[68] The establishment of a fixed center is only a sign of inertia. In this connection, Gadamer remarks that "language is not its elaborate conventionalism, nor the burden of pre-schematization with which it loads us, but the generative and creative power increasingly to make this whole fluid."[69] Tradition should be understood not as a repository of privileged meanings but as a reservoir of topoi or points of departure for an endless process of improvisation and experiment.

What matters is the present act of rhetorical invention. The meaning of Being can therefore be displaced onto the occasion of speech, which becomes, in Blake's phrase, "one infinite plane," while the object of knowledge, the world that can be *known,* is "a vortex pass'd already," and no longer a palpably immediate reality. If knowledge is to be more than the simple amplification of tradition, then the self must always be willing to forgo its present understanding of reality in order to return to the world in its indifferent specificity. Only by opening itself to an encounter with the unforeseen spontaneity of the immediate occasion can the interpreting subject produce an interpretation saved from obsessive repetition. Instead of forgetting the irreducibly specific character of the present, the interpreter must constantly recur from a self-serving fantasy of meaning to the suchness beyond and enclosing the realm of fantasized satisfaction. Desiring energy would then be invested in newly improvised interpretations rather than a supposedly permanent truth. The mind would involve itself in an endless process of *beginning again,* continually adjusting its trajectory as new points of departure suggest themselves in the course of interpretive invention.

This insistence on undermining the self's ineradicable desire for a center that will allow the senses to rest must be qualified in one crucial respect. If the modern suspicion of the desire for fullness, unity, and closure can never finally dispel that desire, if our deep yearning for continuity, coherence, and identity cannot be abol-

68. Said, p. 376.
69. Gadamer, p. 498.

ished, then at least intermittent reliance on received knowledge should not be forbidden. Even Jacques Lacan's poetically half-comprehensible "mid-speak"—a form of utterance close to the unpredictable spontaneity of unconscious mental life—is punctuated by moments of sense. And Nietzsche too recognizes the value of tradition when he argues that the inventor's "freedom from convictions" does not preclude his or her having any: "Conviction as a *means:* many things are attained only by means of a conviction."[70] If unreflecting attachment to a consoling image of human being inhibits innovative thought, the contrary extreme is equally undesirable. The unqualified affirmation of perpetual novelty condemns us to eternal frustration. Endless interpretation has as its obverse an ascetic refusal to enjoy the undeniable pleasure of even a provisional totality.

Tradition is not, then, merely that which must be overcome. Pertinent here is Gerald Bruns's suggestion that we need to counterpoint what he terms a "Romantic" insistence on originality with a "rhetorical" conception of invention as "the art of finding something to say among everything that has already been said":

> Rhetorical improvisation is related to embellishment and ornamentation or adornment; it is an art of doing something to what has already been done. In music and poetry it is a way of exceeding what is written by working between the lines or in the margin, or by using the text as a point of departure or pretext or as a program of intervals. Improvisation in this case is not an art of free origination; it begins instead with what is received, which it then proceeds to color, amplify, alter, or fulfill, never to abolish or forget.[71]

Bruns envisions a form of improvisation that does not dogmatically repeat but rather adopts a playfully deferential attitude toward the past. For Bruns, tradition is not merely a fetish, indicating the inability of human beings to confront experience free from illusions of predictable sameness, but is also a talisman, providing the oppor-

70. Friedrich Nietzsche, "The Antichrist," in *The Portable Nietzsche,* trans. Walter Kaufmann, rev. ed. (London: Chatto and Windus, 1971), p. 638.

71. Gerald L. Bruns, *Inventions: Writing, Textuality, and Understanding in Literary History* (New Haven: Yale University Press, 1982), p. 147.

tunity to experience moments of a perfected existence. Just as radical innovation offers release from the burden of the past, so the consolations of tradition offer release from the burden of invention. If we would resolve the antithesis of dogmatism and creativity in favor of the latter term, we would resolve the complementary antithesis of satisfaction and self-denial in favor of the former. An ideal style of personal comportment would thus involve both pleasure and transcendence, or better, *pleasure in transcendence.* By temporarily satisfying the need for consoling wholeness, the experience of repetition would in fact actually facilitate the future production of difference.

But the redemptive pleasure that issues from tradition can be distinguished from a death-driven desire for finality only so long as repetition is recognized as such. Beautifully cohesive form must foreground its occasional nature lest the centering-effect it affords henceforth commit the mind to a fixed itinerary. An experience of closure that successfully disguised its subjection to history would effectively preclude future occasions of sublime innovation. This need for a rhetorical awareness of the occasionality of utterance, for an ongoing skeptical self-regard, applies not only to those moments when tradition is reaffirmed but to those other times when the mind seeks radically to begin again. Even the most inventive activity of making sense must continually serve notice that "this" production of meaning is taking place at "this" particular time. Speculative thought will thus self-consciously affirm its own rhetorical artifice, with the consequent implication, according to Edward Said, that knowledge be understood "as pathos, invention, imagination—with their pitfalls unobscured."[72] The "flowers of rhetoric" are not simply stylistic adornment but comprise the substance of argumentation itself. "Truth" is a rhetorical effect, a product of the eloquence of the individual speaker. Persuasion depends as much on the pathetic force of speech—on the ability to evoke an aesthetic and ethical response in an audience—as it does on scientific correspondence with a reality external to language. The meaning of Being is always inflected, finally, by a self-pleasuring desire to interpret the world as

72. Said, p. 368.

a repetition of personal tradition. But as the example of Barthes's *S/Z* shows, even an open avowal of the reality of persuasion can provide no perfect safeguard against self-deception. True to his ancient reputation, Hermes remains a trickster, liar, and thief. The personal character traditionally attributed to the patron deity of interpretation is meant to call attention to the delusory nature of every positive assertion. The contemporary renewal of a Gorgianic conception of Being and Knowledge—according to which desire's protean mobility gives rise to an infinite succession of provisionally valid interpretations—will therefore entail the view that Communication depends on a never-ending struggle against self-serving versions of reality. Only by seeking constantly to uncover its own rhetorical artifice can speculative thought retain the ability to adapt to changing circumstance. By turns serious and ludic, philosophical and literary, the activity of invention must be founded on the recognition that originality is only the instauration of a new commonplace.

The following section of *Kaironomia* will mark the passage from theoretical exposition to practical demonstration; a movement, in other words, from Argument to Proof. At the same time, it must be remembered that the historically relative character of knowledge insures that every theory will eventually be understood as a practice in spite of itself. Alternatively, every interpretive practice is secretly nostalgic for the timeless validity of theory. The middle voice remains the middle voice only so long as it loses itself in its own process. As soon as it recognizes itself as middle, it has already become the reflexive voice. Jacques Lacan puts it this way in his version of the Cartesian *cogito:* "I think where I am not, therefore I am where I do not think."[73]

73. Jacques Lacan, "The Agency of the Letter in the Unconscious," in *Ecrits: A Selection,* trans. Alan Sheridan (New York: Norton, 1977), p. 166.

PART THREE

Interpretive Mobility and the Sublimation of Desire

We speak, it is true, of "making" a joke; but we are aware that when we do so our behavior is different from what it is when we make a judgment or make an objection. A joke has quite outstandingly the characteristic of being a notion that has occurred to us "involuntarily." . . . The thought which, with the intention of constructing a joke, plunges into the unconscious, is merely seeking there for the ancient dwelling-place of its former play with words. . . . There is no way of deciding at a first glance whether any element that admits of a contrary is present in the dream-thoughts as a positive or as a negative. I must state emphatically that this fact has not up to now met with any recognition. But it seems to point to an important characteristic of unconscious thinking, in which in all probability no process that resembles "judging" occurs. . . . But besides this, it must not be forgotten that the nonsense in a joke is an end in itself, since the intention of recovering the old pleasure in nonsense is among the joke-work's motives.

—FREUD, *Jokes and Their Relation to the Unconscious*

In learning we tend to hold to one or the other side of relative distinctions which hide the truth, but in *tao* all distinctions are not taken seriously and, in fact, are to be ignored so that one would not hold to one side of a relative distinction against the other. The state of nondistinctness or the state where no distinction commands attention is described by Lao Tzu as that of *voidness* . . . and/or the state of *tranquility*. . . . It is a state where all things are *allowed* to thrive and one can see a natural return of them to their origin—the *tao*.

—CHUNG-YING CHENG, "Nature and Function of Scepticism in Chinese Philosophy"

The Text is experienced only in an activity of production. . . . goes to the limit of the rules of enunciation. . . . *behind* the limit of the *doxa*. . . . is always *paradoxical*. . . . practices the infinite deferment of the signified. . . . is realized . . . according to a serial movement of disconnections, overlappings, variations. . . . is that space where no language has a hold over any other, where languages circulate.

—ROLAND BARTHES, "From Work to Text"

It may happen that some listeners, missing the usual structures (or landmarks) by which they are used to orient themselves may experience some initial difficulties in actually perceiving the music. However, when it becomes apparent that nothing "happens" in the usual sense, but that, instead, the gradual accretion of musical material can and does serve as the basis of the listener's attention, then he can perhaps discover another mode of listening—one in which neither memory nor anticipation (the usual psychological devices of programmatic music, whether Baroque, Classical, Romantic or Modernistic) have any place in sustaining the texture, quality, or reality of the musical experience.

—PHILIP GLASS, liner notes to *Music in Twelve Parts, Parts 1 & 2*

James Joyce defines the relationship between tradition and the present as an interpretive dynamic in the following passage from *Finnegans Wake:* "Imagine the twelve deaferended dumbbawls of the whowl abovebeugled to be the contonuation through regeneration of the urutteration of the word in pregross" (284.18–22).[1] Alluding to the twelve essays collected in *Our Exagmination Round His Factification for Incamination of Work in Progress,* Joyce suggests that in their theoretically disinterested *examination* of *Finnegans Wake,* his interpreters simultaneously *exaggerate* its meaning, or construe the text for their own purposes.[2] That is, to the extent that they are *deafened* by the *howl above bellowed,* the twelve *different* essayists must fail to hear what Joyce is actually saying in the *Wake.* Consequently, when they begin to *bawl* in turn, the meanings they attribute to *Finnegans Wake* are not so much supplementary elucidation as they are original acts of invention taking Joyce's text as their point of departure. Just as the *Wake* is a *factual falsification,* an interpretively biased reworking of traditions preceding it, so *Our Exagmination*'s *thundering continuation* of Joyce's attempt to recapitulate the origins of human culture itself constitutes an authentically new beginning.

The source for Joyce's choice of *thundering* as a term to describe

1. This and subsequent quotations from *Finnegans Wake* are identified by page and line number. Where quotations are drawn from the marginalia and footnotes of Chapter 10, "L" indicates left and "R" right margin; "F" with a number beside it refers to a footnote. There are no chapter titles in *Finnegans Wake;* those appearing in this discussion are given for convenience' sake. In what follows, the source for most glosses of the *Wake* is Roland McHugh's *Annotations to Finnegans Wake* (Baltimore: Johns Hopkins University Press, 1980).

2. Samuel Beckett et al., *Our Exagmination Round His Factification for Incamination of Work in Progress* (1929; rpt. New York: New Directions, 1962). During the years of its composition, *Finnegans Wake* was known as *Work in Progress.*

the act of interpretation is Giambattista Vico's *New Science*.[3] In *The New Science,* Vico describes the evolution of civilization in terms of a principle of cyclical recurrence: societies pass from a theocratic stage of social organization through an aristocratic to a democratic one and finally to social chaos, at which point the cycle repeats itself. The advent of each revolution of this cycle of cultural history is heralded by a divine thundering. Thus, for Vico, to thunder is to pronounce with inaugural force. But according to his cyclical conception of historical process, each cycle newly inaugurated by divine thunder is nevertheless a transformation (or repetition) of every preceding cycle, so that thundering can also be understood to represent a dissembling of originality. In this sense, the writers of *Our Exagmination* could be said to have as their purpose an *in camera incrimination* of Joyce's *Work in Progress*. Challenging its claim to originality by supplying explanatory contexts missing from the *Wake* itself, they would insist that even as it breaks with the past, *Finnegans Wake* is a product of tradition. As Joyce himself concedes, *Finnegans Wake* is "the last word in stolentelling" (424.35). Meanwhile, Joyce can remind these *dumbbells* of their own lack of originality, of the derivative nature of their own interpretive standpoints. Instead of genuinely inaugurating a wholly new *logos* of Being, when they regenerate the Word that animates reality, they are merely beginning *again*. Every act of invention is caught within a logic of repetition and difference; in the language of *Finnegans Wake,* "the same renew" (226.17).

For Joyce, then, our interpretation of the world unfolds as an interminable struggle between an impulse to repeat and the desire to make a difference; a struggle in which the tendency to sterile self-repetition can (temporarily) be overcome in a mutually transforming encounter of personal with alien tradition. According to Edward Said, Vico's *New Science* provides an example of such an interpretive practice, repeatedly renewing its argumentative resources by taking whatever tradition is conveniently at hand as *topos* for fur-

3. Giambattista Vico, *New Science,* revised trans. of the 3d ed. (1744) by Thomas Goddard Bergin and Max Harold Fisch (Ithaca, N.Y.: Cornell University Press, 1968). In the present passage, "thundering" can be heard in "contonuation": *tono* is Latin for "thunder forth."

ther invention. Vico's "topical method" "is everywhere to amass evidence by correlation, complementarity, and adjacency. Although his desire is to locate a primeval beginning, a line of direct filiality, the material testimony of language and his learning restrain his desire, engaging it instead with the susceptibility of language to divination and poetry. A distant and irrecoverable origin is not yearned after fruitlessly because the mind can reexperience its making power by forging novel connections."[4] Vico thus extends the possible ways of making sense by subordinating interpretive closure to an ongoing interpretive practice. Leo Bersani has suggested that such an ideal practice of rhetorical invention necessarily operates according to a principle of juxtaposition, a sort of conceptual collage. That is, successive encounters with the world both "bring a certain coherence" to the activity of invention when it forges novel connections, and "accelerate its disruptive interpretive mobility" by emphasizing the provisional character of any invented equilibrium.[5] The collision of divergent traditions—"A collideorscape!" (143.28)—acts as a continual provocation to invention in the present.

The following Proof for the Argument developed in the preceding section of *Kaironomia* will therefore adopt as its procedure just such a conceptual collage. It will attempt to demonstrate an ideally speculative interpretive practice by taking the juxtaposition of heterogeneous traditions—"a world of differents" (417.10)—as method for renewing the possibilities of invention. Sequential exposition thus periodically will be fragmented in order to multiply the occasions of thought. Even as it achieves a certain coherence, the discussion will cut itself short, and begin again on the basis of a new point of departure: "one continuous present tense" (186.1) interpretive practice staging a breakout whenever any interpretation of the world threatens to become definitive. This self-consciously dramatized meditation on the nature of invention will subordinate meaning to occasion, theory to practice, by resisting the demand for conceptual resolution. Successive traditions invoked in the course of

4. Said, p. 352.
5. Bersani, *Baudelaire and Freud*, p. 7.

its unfolding will augment what has gone before without being absorbed by it. The sense of a meaningful whole will depend upon the perception of analogy among complementary perspectives rather than identity or perfect synonymy. Acceding finally to no magisterial discourse, competing interpretive standpoints will instead engage in a dialogue destabilizing every point of arrest to produce a "proteiform graph . . . a polyhedron of scripture" (107.8), in other words, a *shapeshifting* and *many-sided* activity of invention.

If consciousness occurs as a rupture in the routine order of things, a purely transitory state of wakefulness, this demonstration of interpretive mobility will nevertheless seek at every moment to comply with the imperative concealed in the title of Joyce's *Work in Progress:* "Finn-agains (that is, *everyone*), wake!" Joyce's injunction to waken from a nightmare of unthinking repetition, or history as the production of meaning according to a fixed paradigm, can be understood as the inspiration for this effort to be engaged in an activity of invention that is always beginning again. Taking *Finnegans Wake* as point of departure, the ensuing Proof will set in motion an interpretive dynamic based on conflict, modification, transformation of the traditions caught up in it: a movement of thought whose direction will shift with every new tradition encountered in the course of its exposition: an improvisational theater of ideas.

❧

In the essay he wrote for *Our Exagmination,* Samuel Beckett argues that for Joyce the history of thought unfolds as a succession of transformations of earlier stages, so that more complex forms evolve from simpler ones in a sequence of metamorphic changes. Joyce's project in the *Wake* is to retrace the steps of this process back to its most primitive level, a transit leading backward from modern philosophic speculation to the mythological thought of prehistory. In his investigation of the origins of consciousness in mythological thought, Joyce seeks to grasp the primitive "making" power of the mind in its historical specificity. That is, Joyce maintains that myths are an archaic and rudimentary attempt to orient the mind with respect to concrete lived experience, rather than ornamented statements of some philosophic meaning.

As Beckett explains, myths are not originally "an allegorical expression of general philosophical maxims . . . but a historical statement of fact, of actual contemporary phenomena, actual in the sense that they were created out of necessity by primitive minds, and firmly believed." Myths must therefore be read *literally,* for they have their basis in material reality: "The root of any word whatsoever can be traced back to some pre-lingual symbol. . . . It is the child's mind over again. The child extends the names of the first familiar objects to other strange objects in which he is conscious of some analogy."[6] In its earliest attempts to make sense of its experience, the mind thus has recourse to sensible reality for the symbols that will eventually issue, after much analogical elaboration, in a realm of philosophical speculation no longer dependent on the concrete. Mythological thought improvises the meaning of the world with whatever natural facts come readily to hand. Moreover, according to Joyce, at the very beginning of thought the primitive mind takes as its first point of departure the most palpable of natural facts, the fact of its own body. The earliest symbols with which the mind seeks to comprehend reality, symbols that determine the subsequent course of cultural evolution, are rooted in immediate physical experience—in the primary bodily processes of ingestion, excretion, sexual arousal, and orgasm. Thus, according to a creation myth repeatedly invoked in *Finnegans Wake,* the beginning of the world took place as an act of divine defecation, the "patrilinear plop" (279.4) of our thundering forefather, "farternoiser" (530.36).

Preoccupied with the question of origins, *Finnegans Wake* would answer this question: "What then agentlike brought about that tragoady thundersday this municipal sin business?" (5.13–14). Alluding to the origin of *tragic drama* in the farcial *goat-dance* annually performed in ancient Athens at the festival of Dionysus, Joyce asks: what caused the tragic farce of history? The *Wake*'s answer to this question might be paraphrased as follows: in the earliest attempts to make sense, the distinction between matter and spirit, body and mind, nature and culture has not yet hardened into an ineradicable antithesis. The paired terms still commingle in something like their

6. Beckett, "Dante . . . Bruno. Vico . . . Joyce" in *Our Exagmination,* p. 11.

original state of undifferentiated unity.[7] The meaning of the world is intimately connected with the occasion of invention. But the opposition of matter and spirit, invented by degrees, soon acquires an inherent character. In the gradual elaboration of the first bodily metaphors, the origin of thought in immediate physical experience is willfully forgotten. Separating mind from body and culture from nature, the spiritual realm declares its independence from the material flux of the world. The mind would thus gain access to a timeless space of eternal meanings by severing every connection with the concrete occasion of thought.

But this willful forgetting of the invented nature of the founding categories of thought merely insures that cultural history will unfold as the mystified repetition of its earliest points of departure. The mind's most primitive attempts to understand the world in terms of its own bodily processes continue to determine the course of mental evolution. The myth of anal creation, for instance, once its contingent basis in physiological sensation has been denied, becomes a permanent axiom of thought, the prototype for subsequent notions of originality and authorship. Rather than inventing on the spur of the moment, then, the mind relentlessly repeats its earliest productions: "All that has been done has yet to be done and done again" (194.10). Forgetting the origin of thought in metaphors derived from the body, it loses the ability to make a difference because it no longer knows what it has to differ from. In other words, a tradition whose provisional character has been forgotten precludes an innovative response to the present occasion. Yet in foreclosing the possibility of genuine invention, the mind remains completely faithful to its original intention. The desire to create a spiritual realm beyond the vagaries of immediate physical experience can already be detected in those early inventions closest to concrete actuality. Thus, the myth of anal creation, of creating the world out of one's own substance, can be understood as a primitive expression of the mind's

7. John B. Vickery makes this point in his essay "*Finnegans Wake* and Sexual Metamorphosis," *Contemporary Literature* 13, no. 2 (Spring 1972): 213–42; for Joyce, the earliest human impulse "is to identify the sacred and the unclean, the holy and the evil, the divine and the sexual" (p. 230).

desire to author the shape of its experience without being subject to historical process.

Even in its earliest inventions, then, the mind denies its dependence on the occasion of thought. Insisting that its first acts of invention are definitive, it repeats them in every context. Cultural history can therefore be regarded, Joyce suggests, as a tragic farce—"the farce of dustiny" (162.2). But if history is a farce, "farce" provides Joyce with a strategy for overcoming history. In Norman O. Brown's words, "farce is the mode of consciousness in which a people take leave of their history."[8] Understood as a derisive rejection of tradition and authority, farce supplies an antidote to the desire for access to a transtemporal realm of permanent truth. It promises a state of wakefulness able to pierce the veil of every delusive tradition whose "unchanging Word is sacred" (167.28). The spirit of farce finds its earliest expression in certain ancient festivals, such as the Roman Saturnalia, during which the world is "turned upside down," when the traditions responsible for the customary coherence of social reality are temporarily suspended. Prominent at such festivals, according to Frazer's *Golden Bough,* is "a burlesque figure personifying the festive season which after a short career of glory and dissipation is publicly shot, burnt, or otherwise destroyed to the feigned grief or genuine delight of the populace."[9] The demise of just such a figure is the occasion for *Finnegans Wake.* This wake, Joyce says, is not so much a funeral as a "funferal" (120.10).

In *Finnegans Wake,* the "Festy King" (85.23) whose demise provokes general celebration is precisely the original thundering patriarch whose myth of anal creation forms the basis of every subsequent claim to an authority transcending particular time and place. According to the "tales all tolled" (275.24) in the *Wake,* the entire history of civilization is but a variation on the career of this

8. Norman O. Brown, *Closing Time* (New York: Random House, 1973), p. 52.

9. James Frazer, *The New Golden Bough,* ed. T. H. Gaster (New York: Criterion, 1959), p. 644. In this connection, see also C. L. Barber's analysis of saturnalian license and festivity as a form of cultural renewal in *Shakespeare's Festive Comedy* (Princeton: Princeton University Press, 1959).

first Lord of Misrule. At the beginning a tribal chieftain, he creates a pyramidal social order with himself at the apex as absolute ruler over the other members of his clan—"the greater the patrarc the griefer the pinch" (269.24–25). But he cannot permanently maintain this privileged position as master of social reality. Even as he creates a social order over which to rule, he scatters the seeds of his undoing: eventually, his sons conspire against him, murder and ritually consume the sacred king in order to acquire his potency—"bolt the thor" (279.39)—and then themselves engage in a struggle for supremacy that ensues in a similar cycle of dynastic rise and fall. Margot Norris observes that "the various actions of *Finnegans Wake* precisely portray competition for coveted positions, notably the role of king, father, and subject, rather than object."[10] Cultural history is determined from the beginning by a fantasy of perfect mastery in which historical process is subordinated to the unchanging law of the father.

The thunderer's desire to arrest historical process manifests itself on several levels. First of all, it shapes his understanding of the world, the typical meanings he attributes to reality. According to the myth he invents about the origin of the universe, "primal made alter in garden of Idem" (263.20–21). That is, *one* made *two* out of *Sameness,* or creation not as the temporary conjunction of biologically irreducible beings but as defecation, in other words, the narcissistic amplification of personal tradition. In another version of the same story, the unsettling implications of sexual reproduction (unsettling, that is, to fantasies of originality) are obscured when the birth of the world is imaged in an incestuous union of father with daughter: "Maker mates with made (oh my!)" (261.8). The thunderer thus consistently expresses his desire to preside over reality. He is in fact obsessed with occupying a central, commanding position, an obsession that subsequently determines the character of social life under his rule: "Primus was a santryman and drilled all decent people" (14.13). Primas, the original lawgiver, was a *sanitary*

10. Margot Norris, *The Decentered Universe of Finnegans Wake* (Baltimore: Johns Hopkins University Press, 1976), p. 30. For a discussion of the role ritual cannibalism plays in the transmission of patriarchal authority, see Freud's *Totem and Taboo,* in *SE,* 13:140f.

sentry who sought to check the indiscriminate flow of social intercourse, to establish a sociopolitical hierarchy in which every relationship is controlled and regulated by patriarchal authority. The law of the father depends on repression, on repression of the fact of difference and the self's inevitably relative position; repression of desire, the will-to-invent, the possibility of becoming other; repression of any movement, finally, that threatens to stray from a fixed itinerary. As Richard Ellmann has pointed out, Joyce shared Blake's view that "the real enemy was Nobodaddy, the spirit of restraint."[11]

Joyce will therefore seek in *Finnegans Wake* to lay bare the delusory character of the thunderer's pretensions to permanent authority. The farce of *Finnegans Wake* resides precisely in the *Wake*'s continuous ridicule of every attempt to stop history. According to Joyce, this "fishabed ghoatstory" (51.13) is a tale about a sacramental *fish* who is also an old *goat* or satyr. The thundering patriarch Humphrey Chimpden Earwicker can never finally transcend his material nature: his first two names call attention both to the line of descent he shares with other primates (*chimp-den*) and to his propensity for sexual indiscretions (*hump-free*), his inability to achieve a state of pure spirituality beyond the intrusions of carnal desire. The thunderer's fantasy of spirituality has in fact an erotic motive: "Ever read of that greatgrand landfather of our visionbuilders, Baaboo, the bourgeoismeister who thought to touch both himmels at the punt of his risen stiffstaff and how wishywashy sank the waters of his thought?" (191.34–192.1). In other words, this patriarchal *baboon*'s *tradition-bound* thought can be explained in terms of the desire for permanent pleasure signified by his "risen stiffstaff." Even as he aspires to touch the *heavens,* his visionary quest has its basis in erotic experience.

The thunderer's career unfolds in this sense as a succession of rationalizations and self-deceptions, a continually renewed effort to avoid recognizing the material basis of spiritual aspiration. His thundering is equally a stuttering, a guilty hesitation ("He was hesitency carried to excelcism" [82.30–31]) in which he inadvertently

11. Richard Ellmann, *James Joyce* (New York: Oxford University Press, 1959), p. 26.

discloses his submission to desire, his relative moment in the stream of history. Joyce's investigation into the origins of consciousness can thus be understood to have as its purpose an exposé of our thundering forefather's fraudulent proclamations of originality. If "the sib-speeches of all mankind have foliated (earth seizing them!) from the root of some funner's stotter" (96.30–31), then by celebrating the demise of the first father to imagine the world as an extension of himself, Joyce inaugurates a practice of abolishing every tradition that denies its historically contingent character.

Consider, for instance, the following phrase: "Ainsoph, this upright one with that noughty besighed him zeroine" (261.23–24). The first reference Joyce makes here is to the Cabalistic myth of Ainsoph, the spiritual author of reality, whose *ten* emanations sustain the material universe. But this assertion of the spiritual realm's ontological priority is immediately challenged by a scatological level of meaning: this *upright one* with that *noughty zero* can be read as an allusion to the coupling of phallus with vagina.[12] Joyce thus implies that the timeless instant of sexual coming provides a model for subsequent notions of eternity. He then introduces a third level of meaning that records the moment when this initial identity of sacred and profane is transformed into an antithesis. That is, the original erring thunderer obscures the physiological basis of eternity by elaborating a myth according to which sexual union represents the seduction of male spirituality by female materiality: this *morally upright* one whose downfall has been caused by that *naughty heroine*. A self-serving myth of the fall thus enables the thunderer to acknowledge his subjection to historical process without sacrificing his illusion of autonomy and self-possession. Taken together, the three levels of meaning describe the manner in which the spiritual asserts its independence from material determinations even as it is derived from immediate physical experience: the "CONSTITUTION OF THE CONSTITUTIONABLE AS CONSTITUTIONAL" (261.R).

By restoring the priority of physiological reality, Joyce hopes to discredit "the whole faustian fustian" (292.22). He would repudiate

12. Margaret Solomon explores the scatological dimensions of *Finnegans Wake* at length in *Eternal Geomater: The Sexual Universe of Finnegans Wake* (Carbondale: Southern Illinois University Press, 1969).

every thundering proclamation of permanent authority by returning thought to its original bodily metaphors: "Phall if you but will, rise you must: and none so soon either shall the pharce for the nunce come to a setdown secular phoenish" (4.17). From the standpoint of *secular farce,* the Viconian cycle of cultural epochs, the Spirit's progress through History, is merely a disguised repetition of the *phoenixlike* rise and fall of the father's *phallus.* For Joyce as for Nietzsche, the history of consciousness unfolds as a "misunderstanding of the body," a confusion, so to speak, of anatomy with metaphysics. In the language of the *Wake,* "The inception and the descent and the endswell of Man is . . . wrapped in obscenity" (150.30–31). Ideally, the discovery of a scatological basis for every tradition—or as Joyce puts it, "farcing gutterish" (518.25)—would free the mind to invent in response to the present occasion rather than according to the dictates of the past. Hence this interpretation of the title of Beckett's essay "Dante . . . Bruno. Vico . . . Joyce": just as Bruno's recognition of the infinity of interpretive standpoints contradicts Dante's finite hierarchical world view, so Joyce—that "misbelieving feacemaker" (301.4)—deploys a sustained moment of farcical derision to undo the inevitability, in Vico's view, of endless historical repetition—"o peace a farce" (14.14).

❧

The term "sublimation" was first used in psychoanalytic discourse by Freud to theorize a material basis for the life of the spirit, a bodily origin for the traditions that give meaning to the world. According to Freud, desire can exchange its original aim of immediate physical pleasure for the more rarefied pleasures of civilized life on condition that cultural tradition metaphorically reproduce a bodily satisfaction which remains the primary focus of desiring interest. The sublimation of desire, an appropriation of instinctual energy providing the impetus for civilization's development, depends on retaining desire's earliest forms of gratification in figurative guise. In other words, the erogenous zones of the body from which pleasure is first obtained continue to orient the course of cultural history. For Jean Laplanche, however, the history of consciousness is a process of transformation whose beginning does not constitute a privileged point of origin. Contradicting Freud, Laplanche maintains the sublimation of desire does not merely operate according to a principle of repetition. Laplanche has in fact taken the topic of sublimation as point of departure for speculation concerning an ideally circumstantial or improvisational activity of invention.

According to Laplanche, desire is born in the organism whenever a stimulus is strong enough to destabilize a preexisting state of equilibrium: "There is the capacity in human being . . . of ceaselessly creating . . . the sexual from every sort of exterior shock."[13] Any intrusion from the world that disrupts tradition or habit, any eruption of the sublime is capable of arousing the subject's desiring interest. The fundamental emptiness defining all of us precisely as *desiring* entities insures that the prospect of becoming other will inspire an effort to assimilate difference to personal tradition in order, for the moment, to fill up the lack in Being. As Laplanche says, "Newly formed sexuality seems able to take as its point of departure *absolutely anything*."[14] The movement of desire from erotic friction to pleasurable release thus applies not only at the mundane level of physical pleasure but is equally descriptive of the workings of consciousness. Cultural traditions are not derived but are themselves primary forms of desiring experience. Instead of endlessly memorializing the body's erogenous zones, consciousness can therefore satisfy desire by producing novel inventions in response to changing circumstance. The ideal sublimation of desire would unfold as a continuous "passage of the instinctual into the life of the self," a mental progress in which the subject's intention is always inspired by its immediate situation.[15] For Laplanche, sublimation signifies the possibility of an "open symbolization" based not on a "closed logic" of linear continuity and univocal sense but on an "ambivalent, ambiguous" one that permanently resists interpretive closure.[16] Understood as the metamorphosis of desire in culture, it

13. Jean Laplanche, *Problématiques III: La sublimation* (Paris: Presses Universitaires de France, 1980), p. 111 (my translation).

14. Laplanche, *Life and Death in Psychoanalysis*, p. 101.

15. Laplanche, *Problématiques III*, p. 194; in this connection, Laplanche suggests that Freud's famous phrase "Where It (or Id) was, there the Self shall be" can be understood to formulate the protean mobility of such a movement of the intellect.

16. Ibid., p. 126.

implies a practice of invention in which the mind would always be ready to modify its interpretation of the world as new and unprecedented desires emerge into consciousness.

But Freud would insist, in opposition to this optimistic view of sublimation, that the eruption of instinctual energy into the life of the self is a *traumatic* occurrence: the self is wounded by desire. Desire is experienced as a threatening intrusion, an influx of energy throwing the organism into a state of panic. The traumatic eruption of desire introduces dissonance into a harmonious whole. It threatens to overturn the self's existing system of meanings. As Melanie Klein remarks, the "depressive anxiety of disintegration," a terrifying experience of ontological insecurity, is a crucial motive for the sublimation of desire.[17] Sublimation can be understood to have as its purpose the recovery of a feeling of untroubled calm. The self would in fact achieve a composure so perfect that it would never again be vulnerable to desire's disruptive impact. According to Norman O. Brown, "The peculiar structure of the human ego results from its incapacity to accept . . . the supreme reality of death and separation," the reality, that is, of the self's subjection to historical process, its potential for becoming different from itself. Instead of an appropriation of instinctual energy newly evoked by the self's surroundings, sublimation unfolds as an "infantile flight from death," a denial of the reality of desire, the impermanence of every satisfaction.[18]

The mind responds to the birth of desire by imagining desire's definitive end. Its earliest experiences of instinctual excitation are so traumatic that it seeks to arrest the movement of desire once and for all. Thus, the child produces oral fantasies of consuming the world, anal fantasies of creating the world, phallic fantasies of presiding over the world, fantasies of satiation, authority, and preeminence that lay the foundation for the characteristic projects of the mature self. Human beings, Freud suggests, have a limited capacity for invention, a capacity normally exhausted in childhood. Forever re-

17. Melanie Klein, *Contributions to Psychoanalysis, 1921–1945* (London: Hogarth Press, 1948), p. 290.

18. Norman O. Brown, *Life against Death: The Psychoanalytical Meaning of History* (Middletown, Conn.: Wesleyan University Press, 1959), pp. 159, 155.

luctant to invent, the mind intends its earliest inventions to be conclusive. The mental evolution of the individual is therefore permanently inhibited by a dogmatic attachment to the first fantasies produced in order to placate desire. As Brown puts it, cultural life is shaped by "unconscious oral, anal, and genital fantasies of return to the maternal womb."[19] The subject's quest for an ideal condition of nondesiring rest, for a "narcissistically self-contained and self-replenishing immortal body," thus causes it to retreat into a realm of imaginary satisfactions. Turning away from reality, it seeks refuge in a fantasy of self-possession. So that instead of authentic transactions with the world, there is the illusion, in Brown's words, of "having the world within the self."[20]

The "self," moreover, is itself invented precisely in order to thwart the disquieting emergence of desire. According to Jacques Lacan, the subject first conceives of an unchanging personal identity as consolation for the discovery of its own desiring nature. That is, the ego is constituted for the first time as an enduring pole of identity during the so-called mirror stage when the infant recognizes its own reflection in a mirror. This recognition serves as an important purpose: the infant compensates for its lack of motor coordination by interpreting its reflection as an autonomous and controllable self. The infant's "jubilant assumption of its specular image," a feat that situates the self "before its social determination in a line of fiction that will only asymptotically rejoin the development of the subject," is thus ruled by an aspiration for mastery or the will-to-power. The self would permanently occupy "the position that no one can really assume because it is imaginary, that of absolute master."[21] It would substitute for the "otherness" of the Other an obedient image of

19. Ibid., p. 291; see also Jeffrey Mehlman, *A Structural Study of Autobiography: Proust, Leiris, Sartre, Lévi-Strauss* (Ithaca: Cornell University Press, 1974). Mehlman shows how the loss of the "original- maternal- object" is never fully accepted: the writers of autobiography experience a desire for unity that has its prototype in the infant at the mother's breast.

20. Brown, *Life against Death*, pp. 293, 162.

21. Jacques Lacan, *Ecrits* (Paris: Editions de Seuil, 1966), pp. 90, 91, 44; Brown makes a similar point when he observes that the "infantile conflict between actual impotence and dreams of omnipotence is . . . the basic theme of the universal history of mankind" (*Life against Death*, p. 25).

itself. Brown remarks, "The vain project of being both Self and Other . . . supplies the human ego with its essential energy."[22] The self presents itself to desire as a perfectly satisfying object of desiring attention. Personal identity is, then, an illusion of finality with which the subject would conceal from itself the contingent, mutable character of human being. Like a fetish or protective charm, the subject brandishes an interiorized image of self-possession at every eruption of desire, hoping thereby to expel an unwanted intruder.[23]

With the constitution of the ego as a fantasy of permanent satisfaction, the subsequent fate of the will-to-invent is sealed. The sublimated activities of cultural life, transpiring within the space of the self, become instruments for preserving the subject's illusory sense of its own autonomy. The quest for knowledge, in other words, has its basis in the narcissism of infancy. Thus, Freud says, children go through a period around the age of three

> which may be called the period of *infantile sexual researches.* So far as we know, the curiosity of children of this age does not awaken spontaneously, but is aroused by the impression made by some important event—by the actual birth of a little brother or sister, or by a fear of it based on external experiences—in which the child sees a danger to his selfish interests. Researches are directed to the question of where babies come from, exactly as if the child were looking for ways and means to avert so undesired an event.[24]

According to Freud, the arrival of a sibling, an event that threatens to displace the child from the center of its universe, inspires the earliest attempt to make sense. The inaugural act of intellectual inquiry is occasioned by the child's anxious need to reaffirm its own preeminent importance. The various theories it produces in order to

22. Brown, *Life against Death,* p. 161.

23. In this connection, see Freud's 1927 essay "Fetishism." Freud remarks there that a fetish is "a substitute for the woman's (the mother's) penis that the little boy once believed in and . . . does not wish to give up" (*SE,* 21:152–53). That is, the fetishist is unable to truly acknowledge the fact of sexual difference because such a recognition would wound his narcissistic belief that the other is merely a repetition of himself. A fetish is thus a device for safeguarding the illusion of centrality.

24. Freud, *Leonardo da Vinci and a Memory of His Childhood, SE,* 11:78.

solve the riddle of existence—whence came I?—consequently express reassuring fantasies of personal authority: "following the lead of the impulses of [its] own sexuality" the child formulates "theories of babies originating from eating, of their being born through the bowels."[25] The child thus begins its investigation into the nature of reality with the comforting illusion that it might itself become sole creator of its own world.

When the child fails finally to put its theories into practice (fails, that is, to author any children of its own), it gradually forgets them. But though forgotten, the early fantasies of perfect mastery are never really given up. The mind remains unconsciously attached to the infantile dream of a permanently pleasurable union with its surroundings. As a result, Freud says, "the free activity of intelligence may be limited for the whole of the subject's lifetime." This inhibition of the intellect may take either of two forms. On the one hand, the mind may continue to invent, but when it does so, it hastens to bring the activity of making sense to a close. In this case, the subject's investigation of the world appears in the form of "compulsive brooding," a form of thought infused "with the pleasure and anxiety that belong to sexual processes proper," impatient, in other words, for an early reduction of desiring tension.[26] Confronted with a problem it cannot immediately resolve in terms of an existing system of meanings, the mind hurriedly forges a semblance of unity which, like every forgery, must in the end be recognized as fraudulent.[27] On the other hand, after failing in its initial attempt to master reality, the mind may abandon invention altogether. It now seeks to consummate its narcissistic aspiration by means of an ideal imposed from without. The child's self-image becomes "the precipitate of the old picture of the parents, the expression of admiration for the perfection which the child then attributed to them."[28] Thus identifying with a parental image of invulnerable authority, the mind ac-

25. Ibid., p. 79.

26. Ibid., p. 80.

27. According to Mehlman, Laplanche makes this point by describing "the ego's characteristic behavior as 'synthesis compulsion': the drive to operate unifications that are ultimately delusory, 'nonsynthetic' "; see Mehlman, p. 182.

28. Freud, *New Introductory Lectures on Psychoanalysis*, *SE*, 22:65.

cedes to the socially ratified narcissism of cultural tradition. The sublimation of desire then becomes the repetition of stereotypical fantasies of satisfaction. In either case, the evolution of consciousness is punctuated by recurrent proclamations of a permanent state of nondesiring stillness.

This continued adherence to the infantile dream of perfect bliss commits the self to a life of anxiety and frustration. The mere fantasy of total satiation can never truly satisfy the self. The phoenix-like rebirth of desire poses an unending threat to the illusion of finality. As Freud observes, the subject's compulsive brooding "never ends, and . . . the intellectual feeling, so much desired, of having found a solution recedes more and more into the distance."[29] The incessant pressure of instinctual energy, continually provoked anew by the self's surroundings, postpones indefinitely a final conclusion. Moreover, even if the subject could withdraw from the world completely, it would still be unable to satisfy its desire. Desire's capacity for endless renewal is not simply a result of the subject's involuntary responsiveness to its environment. Instinctual energy is in part self-generating. From the moment it attaches itself to one point of view, a contrary standpoint emerges as a further possibility. The fantasy of permanent satisfaction is therefore doomed from the start. The self cannot escape its shadow: for every attribute defining the character of personal tradition, there resides in the unconscious a repressed antithesis. In this connection, Brown remarks that humanity "separates the opposites, turns them against each other, and, in Nietzsche's phrase, sets life cutting into life."[30] Unable to abide the reality of tragic dissonance, the self pursues a futile project of harmonious resolution. It refuses to concede the intrinsic heterogeneity of desire, the necessity of multiple stances toward the world. Insisting on unambiguous meaning in the midst of desire's thriving ambivalences, the mind consequently discovers itself both threatened and tempted by forbidden objects of desiring attention.

But an unhappy career of frustrated craving is not, for all that, the subject's inevitable destiny. The troubling mobility of instinctual

29. Freud, *Leonardo da Vinci, SE,* 11:80.
30. Brown, *Life against Death,* p. 84.

energy might itself be said to provide a solution to the problem of insatiable desire. Instead of a panic-stricken quest after perfect bliss, the self could make the most of desire's protean character by seeking satisfaction in the always provisional pleasures of the moment. The possibility of a more flexible, improvisational selfhood can be extrapolated from Freud's account of the "polymorphous perversity" of the sexual instincts before the subject has actively negated their mobility. In the beginning, Freud says, the sexual instincts "are numerous, emanate from a great variety of organic sources, act in the first instance independently of one another, and only achieve a more or less complete synthesis at a late stage."[31] Before the convergence of desire on the self as a unified image of the body, desire lacks a controlling center. The infant's earliest inclination, in Brown's words, is to "explore in indiscriminate and anarchistic fashion all the erotic potentialities of the human body."[32] This original mobility of instinctual energy remains a present reality in the adult unconscious. According to Freud, two principles govern the movement of unconscious desire: condensation and displacement. Freud remarks that "by the process of *condensation* [one idea] may appropriate the whole cathexis of several other ideas." That is, the desiring energy invested in one group of ideas may be transferred to some similar idea able to serve as their metaphorical representative. In condensation, desire moves among elements formally related to one another. It is thus a principle of repetition. Displacement, on the other hand, and here the really productive character of desire shows itself, is a principle of radical difference. By the process of displacement "one idea may surrender to another its whole quota of cathexis" on the basis of mere proximity: the two ideas just happen to be contiguous.[33] Displacement refers to a movement of desire unconstrained by a traditional itinerary. It insures that the student of the unconscious will forever be confronted with an "indefatigable variety of new shapes."

What Freud terms the "primary process" of the unconscious consequently transpires as a ceaseless forming and reforming of

31. Freud, "Instincts and Their Vicissitudes," *SE*, 14:125.
32. Brown, *Life against Death*, p. 27.
33. Freud, "The Unconscious," *SE*, 14:186.

meanings in which momentary condensations of sense are continually deranged by the metonymic activity of displacement. As Leo Bersani observes, it is "fundamental to desire that it should constantly be detaching itself from its objects and finding new representations. . . . To desire is to move to other places."[34] While condensation provides for centers of intelligibility, these centers are never more than provisional because displacement insures an inevitable slippage of meaning. The desiring flux of the unconscious can in this sense be understood to promise the reconstitution of the self as something other than a permanent impediment to desire. By transforming the "secondary process" of the conscious ego into an extension of the id's "primary process," the subject might once more be able to invent. The sublimated desiring mobility of speculative thought would then emulate the exuberant fantasia of the unconscious, becoming, in Laplanche's phrase, a form of "extemporaneous sexuality." No longer an identity, the self would henceforth comprise the activity of continually translating desiring energy into cultural products.

Such an activity of sublimation would unfold as a sequence of contingent moments of interpretive closure. Forsaking the illusion of finality, the subject would deliberately seek out traumatic encounters with unprecedented desire. Laplanche has in fact suggested that a crucial component of inventiveness is "traumatophilia," a desire to be confronted with the irreducibly other. To refuse to be traumatized is to insist the present is nothing but a repetition of the past. Alternatively, the willingness to take traumatic difference as point of departure for the sublimation of desire would ensue in a continuous conversion of experience into knowledge, what Freud calls "the rarest and most perfect" type of sublimation.[35] The investigation impulse would then no longer aspire to a definitive conclusion,

34. Bersani, *Baudelaire and Freud*, p. 61.

35. Freud, *Leonardo da Vinci*, *SE*, 11:80. In the following passage from *The Exile of James Joyce*, trans. S. A. J. Purcell (New York: D. Lewis, 1972), Hélène Cixous defines the Joycean artist precisely as traumatophilic: "The first motive behind the necessity to write and to be an artist is not purely the manifestation of a desire to create for the sake of creating and thereby to experience a divine feeling of omnipotence, but, on the contrary, the creative impulse is caused by a painful

satisfying itself instead with a transitory equilibrium. Although here too, as Freud points out, "the research becomes to some extent compulsive and a substitute for sexual activity; . . . there is no attachment to the original complexes of the infantile sexual research, and the instinct can operate freely in the service of intellectual interest."[36] In such a movement of consciousness, metaphoric condensations of meaning would be disrupted, before they had become rigidly fixed, by metonymic displacements of desiring attention. That is, when it escapes rote repetition, the sublimation of desire continually proceeds, in Melanie Klein's words, "to a sexual symbolic cathexis of further ego-activities and interests," so that even as it repeats itself, it differs from the past.[37] The old thus would be discarded in favor of new thought evoked in response to the subject's shifting surroundings. Instead of assimilating desiring interest to a preexistent structure, the subject would constantly improvise, creating in a fresh eruption of sexuality a new intentional product, a new transaction between the "inside" and the "outside."

In this connection, Laplanche suggests that Freud's account of the acquisition and control of fire as the inaugural act of sublimation, his interpretation of the myth of Prometheus the firebringer admits of another reading. According to Freud, humans acquired the use of fire when they mastered the urge to extinguish its flames by urinating on them: "The first person to renounce this desire and spare the fire was able to carry it off with him and subdue it to his own use. By damping down the fire of his own sexual excitation, he had tamed the natural force of fire. This great cultural conquest was thus the reward for his renunciation of instinct."[38] Prometheus is then a hero of instinctual renunciation who makes possible the

---

encounter with real life, illustrating the hostility of life and its intentions of altering the hero's subjective nature." According to Cixous, the creative act, "at first defensive, becomes offensive later, since it is not sufficient simply to have repelled the aggressor; and fear becomes a desirable possession because it excites the rebellious imagination. Hence, the artist finds it necessary to arouse and provoke the aggression against which he is to write" (pp. 396–97).

36. Freud, *Leonardo da Vinci*, *SE*, 11:80.

37. Klein, p. 115.

38. Freud, *Civilization and Its Discontents*, *SE*, 21:90; see also "The Acquisition and Control of Fire," *SE*, 22:187–93.

emergence of civilization by directing desire away from immediate physical gratification toward a purely mental realm. Desire will henceforth have its principal outlet in the form of ideas and intentions, in other words, in fantasies of satisfaction whose realization is presently deferred. At the very moment that he invents civilization, Prometheus thus condems humanity to a life of frustration.

But since fire is itself a traditional emblem of sexual excitation, then Prometheus's theft of fire from the gods might refer, Laplanche says, to "an appropriation of bliss without limit." The myth of Prometheus has this double meaning: there is a competition or play between two manifestations of desire, a "play between two penises, the penis of water and the penis of fire."[39] Laplanche explores the implications of this dynamic of fire and water by juxtaposing the career of Prometheus with that of Hercules, who rescues Prometheus when Zeus punishes him for his theft. Hercules, Laplanche argues, is undeniably a libidinal hero, associated with an unfettered expense of desire, as when he *floods* the Augean stables. The relationship between the two figures recapitulates the relationship between the primary and secondary processes and in effect defines the space within which an ideal sublimation of desire must situate itself. Thus, where Hercules represents an incoherent desiring frenzy seeking an ultimate discharge, Prometheus stands for a channeling of instinctual energy that renews rather than extinguishes the movement of desire. And where Prometheus stands for a repudiation of present pleasure and consequent retreat into dogmatic fantasy, Hercules would be that intrusion of instinctual energy which bursts through every structure intended to contain its flow: a traumatic eruption of desire enabling a new beginning.

❧

The "Nightlessons" chapter of *Finnegans Wake* (260–308) is arranged in the manner of a school textbook complete with marginalia and footnotes. The chapter contains a Cabalistic account of the world-creating descent of spirit into matter, a history of the rise of civilization according to Vico's theory of cyclical recurrence, and a survey of major departments of knowledge. In "Nightlessons,"

39. Laplanche, *Problématiques III*, pp. 164, 167.

Shem, Shaun, and Isobel, the three children of the primordial parents Humphrey Chimpden Earwicker and Anna Livia Plurabelle, are introduced to the diverse accomplishments of human culture. As they put it themselves at the end of their lessons: "We've had our day at triv and quad and writ our bit as intermidgets. Art, literature, politics, economy, chemistry, humanity, &c" (306.12–15). The children thus receive a liberal arts education, a grounding in the fundamental premises on which civilization is based.

From "the murk of the mythelated in the barrabelowther" we proceed, then, "to the clarience of the childlight in the studiorum upsturts" (266.9–13). Leaving behind the adult world, whose inhabitants are so *intoxicated by myth* they can only repeat the cyclical round of history, we now ascend to the *clarity* of the children's *study,* where these young *upstarts* may yet break the cycle of endless repetition. The twin sons are enjoined, in other words, *not* to emulate their father: "Here, Hengegst and Horsesauce, take your heads out of that taletub! And leave your hinnyhennyhindyou! It's haunted. The chamber. Of errings" (272.17–20). In the allusions he makes here to the writings of Jonathan Swift, Joyce first of all associates Shem and Shaun with the Houyhnhnms of *Gulliver's Travels,* suggesting that in the "clarience" of their "childlight" they resemble those noble *horse*-like creatures whose reason and virtue satirically contrast with the low venality and lust of the humanoid Yahoos (or adults). He then cautions the twins against deferring to *A Tale of a Tub*'s putative author, depicted by Swift as an inmate of Bedlam: "Look at your mad father on his boneshaker fraywhaling round Myriom square" (285.F4), "that wheeze of old windbag, Blusterboss, blowharding about all he didn't do" (273.22–24). Thus forewarned about the delusions of grandeur that accompany the exercise of power, when *Hengest* and *Horsa* (or the Anglo-Saxons) overthrow Arthurian Britain, they will not merely repeat the old king's *errors.*

The children's introduction to cultural tradition can be understood in this sense precisely as a biography of their thundering father Humphrey Chimpden Earwicker in whose career they discover the beginning principles of civilized life: "As we there are where are we are we there from tomtittot to teetootomtotalitarian" (260.1–3).

Cultural history begins with "tomtittot," *a demon whose safety depends on concealing his identity,* in other words, a patriarch whose authority depends on denying his contingent status as a product of history. This inaugural attempt to stop time subsequently becomes the progenitor, by way of "teetootom" or *a throw of the dice,* of modern totalitarianism. The aleatory movement of human history—"Willed without witting, whorled without aimed" (272.4)—nevertheless contains a recurrent theme: the spectacle of successive patriarchs each seeking to remain permanently on top of the social pyramid: "Hispano-Cathayan-Euxine, Castilian-Emeratic-Hebridian, Espanol-Cymric-Helliniky? Rolf the Ganger, Rough the Gangster, not a feature alike and the face the same. . . . [Curse] you for an archetypt!" (263.13–16, 29–30). The children are encouraged to reject this *archetype* of every authority, *architect* of history's futile cycle of dynastic rise and fall.

In order to discredit patriarchal authority, "Nightlessons" reveals the thunderer's guilty secrets to Shem, Shaun, and Isobel, who "gang along to gigglehouse, talking of molniac's manias and missions for mades to scotch the schlang" (289.18–19). That is, the lessons the children learn in this "gigglehouse" of farcical derision concern the thunderer's *megalomania* and his projects for *stamping out* the *snake,* "the glider that gladdened the girl" (271.25–26), precisely the phallus whose ceaseless renaissance reminds him of his own desiring nature: "The phoenix, his pyre, is still flaming away" (265.8–9). Informed of his intrinsic carnality, the thunderer's restive subjects need no longer be intimidated by his pretensions to spiritual authority: "One hath just been areading, hath not one, ya, ya . . . ga, ga . . . da, da, of Sire Jealyous Seizer, that gamely torskmester, with his duo of druidesses in ready money rompers and the tryonforit of Oxthievious, Lapidous and Malthouse Anthemy" (270.29–271.6). The political struggles at the end of Rome's republican era can be reduced to the sexual indiscretions of a *jealous* patriarch seeking exclusive access to "his duo of druidesses in ready money rompers." The thunderer's repression of sexuality does not, then, imply his commitment to an exclusively ascetic ideal. Although Earwicker is a "Salmonson" (297.3) or sacramental *fish* who restricts the free flow of desire within himself when it challenges his illusion of spiritual autonomy and self-possession, he is also a "gamely torskmester" or patriarchal and profane *cod* (as in "codpiece") who would facilitate his own unlimited pleasure by compelling universal obedience to his will, as when "the vaulting feminine libido" is "sternly controlled . . . by the uniform matteroffactness of a meandering male fist" (123.8–10).

With this recounting of the thunderer's sexual activities, the children's quest after the origins of culture arrives at the conjugal bed of "Airyanna and Blowyhart topsirturvy, that royal pair" (275.14–15). Civilization begins with "the marriage of Montan wetting his moll" (260.16–17). The force propelling cultural evolution can be traced to the sexual coupling of private parts: we have all been "blessed by Pointer the Grace's his privates judgements" (289.21–22), by "Lang Wang Wurm, old worbbling goesbelly" (270.F2) whose favorite pastime is "beeswaxing the convulsion box" (261.L). A scatological level of meaning proposes, then, that humans are inescapably pleasure-seeking creatures: "It's tails for toughs and titties for totties and come buckets come bats till deeleet" (272.6–8); the orgasmic union of "the three legged man and the tulippied dewydress" (331.8–9) as the basis of All. In the language of the *Wake,* "life is body love" (271.9).

The children thus discover that questions of knowledge and sexuality intersect: "Here we'll dwell on homiest powers, love at the latch with novices nig and nag" (266.13–15). Political history becomes sexual politics—"Joke and Jilt will have their tilt" (290.F2)—and even algebra acquires a scatological resonance: "How many combinaisies and permutandies can be played on the international surd"?[40] Answer: "If the two antesedents be bissyclitties and the three comeseekwenchers trundletrikes, then, Aysha Lalipat behidden on the footplate, Big Whiggler restant upsittuponable" the number of combinations "presents to us . . . an ottomantic turquo-indaco of pictorial shimmer so long as, gad of the giddy, pictorial summer" (284.12–14, 22–30). That is, during the

40. In this context, "surd" compresses antithetical meanings: on the one hand, it can be read as a rhyming allusion to the thunderer's deluded belief in anal creation; on the other, it is an irrational or nonfinite number whose aleatory, unpredictable character contradicts the notion of an unseen hand controlling the shape of events.

summertime of his reign, when he is at the height of his potency, Earwicker sires an entire society, a *rainbow*'s array of "combinations" on his primordial wife Anna Livia. But once he establishes a social hierarchy, his offspring begin to contest his rule, thereby setting in motion history's endless "permutations": "outraciously enviolated by a mierelin roundtableturning," old King Arthur now appears as "the losed farce on erroroots" and the permutations of "twalegged poneys and threehandled dorkeys . . . brings us a rainborne pamtomomiom" (285.2–3, 12–16), the *pandemonium* of dynastic upheaval. This model of historical development is repeated many times in the course of the children's lessons; here, for instance, in the guise of a geometry problem: "Show that the median, hce che ech, intersecting at royde angles the parilegs of a given obtuse one biscuts both the arcs that are in curveachord behind. Brickbaths. The family umbroglia" (283.32–284.4). In other words, the *rigid* right angle intersection of HCE—Earwicker's initials—with Anna Livia's "parilegs" produces the family *imbroglio*. The tangled confusion of human history becomes comprehensible as a network of *prickpaths*.

There is a second and more important geometry problem in "Nightlessons." The solution to this one involves uncovering the secret of the mother's sexual identity. In the second half of "Nightlessons," Shem and Shaun diagram the "whome" of our "eternal geomater" (296.31–297.1), Anna Livia Plurabelle. Earwicker's twin sons thus conduct an investigation into the mystery of procreation, an investigation that is decisive for the subsequent development of their respective personalities. Their task is, then, to "construct ann aquilittoral dryankle Probe loom!" (286.19–20), an *equilateral triangle,* traditional emblem of the female. The sons complete this assignment by peering under Anna Livia's skirt, "the maidsapron of our A.L.P." (297.11), which when lifted reveals her pubic region, here compared to the alluvial deposit at the mouth of a great river: "the midden wedge of the stream's your muddy old triagonal delta . . . the no niggard spot of her safety vulve, first of all usquiluteral threeingles . . . the constant of fluxion" (297.27–30), a truth unchanged by the stream of time, precisely "her it. Which it whom you see it is her" (298.2–3). Shem and Shaun have just caught their first glimpse of adult sexuality, their first glimpse, that is, of irreducible difference and insatiable desire: "the It with an itch in it, the All every inch of it, the pleasure each will preen her for, the business each was bred to breed by" (268.3–6).

But while Shem welcomes this confirmation of Anna Livia's carnality, Shaun resents being "made vicewise" (286.29). The twin sons' sexual investigation embodies contrary perspectives, one "INGENUOUS," the other "LIBERTINE" (286.R), one associated with Mark Twain's "topsowyer" (299.28), the other with "Hurdleberry Fenn" (297.20). Where Shem—"Mr. Tellibly Divilcult"—regards human being as inseparable from material reality, Shaun—"He, angel that I thought him" (303.F1)—pursues what Freud describes in *The Interpretation of Dreams* as the ego's "ostrich policy" of shutting out whatever threatens its narcissistic self-enclosure.[41] They make, then, "a daintical pair of accomplasses! You, allus for the kunst and me for omething with a handel to it" (295.27–29). Shem insists on *artistic* truth, no matter how unflattering, while Shaun inclines to the self-serving view of an expedient *businessman*. "Nightlessons" (indeed, the whole of *Finnegans Wake*) transpires in "Draumcondra's Dreamcountry" (293.F1). It presents a *dream* vision of cultural evolution as the simultaneous recognition and denial of the *trauma* of sexual difference, of human rootedness in history. The sons' "traumaturgid" (496.24) investigation of the mother is consequently suffused with anal eroticism from the very beginning: "First mull a mugful of mud . . . Anny liffle mud which cometh out of Mam will doob, I guess" (286.31, 287.7–8). Discovery of the mother's role in procreation is obscured by an anal fantasy intended to shore up male narcissism: any little "mud" that comes out of *Man* will do. Even as Shem completes his diagram of Anna Livia's genital anatomy, Shaun misinterprets its significance: "It's the mud-

41. See Freud, *The Interpretation of Dreams, SE,* 5:600. Lacan elaborates on the ego's ostrichlike behavior in his "Seminar on the Purloined Letter"; see *Ecrits,* pp. 19–53. It is interesting to note in this connection the concluding image of Luis Bunuel's 1974 film *The Phantom of Liberty:* precisely, an ostrich. In that film, Bunuel attempts to undo the human propensity for self-imprisonment by converting normative values and assumptions about the world into their opposites. He thus brings to light a background of alternative possibilities against which the seeming inevitability of custom and habit defines itself.

dest thick that was ever heard dump"—sexual reproduction implicitly understood as a derived form of anal creation—"since Eggsmather got smothered in the plap of the pfan" (296.20–22)—since both *mother* and *matter* were repudiated by a thunderer pretending to be the origin of himself: "And he clopped his rude hand to his eacy hitch and he ordurd and his thick spch spck for her to shut up shop, dappy" (23.3–5).[42]

Actually, the prospect of a truly carnal mother elicits a contradictory response from Shaun. Insofar as he identifies with Anna Livia, he spiritualizes her, denying her material nature in order to retain his own sense of transcendence. She thus becomes the "Ensouling Female" who sustains "Agonizing Overman" (302.L). For Shaun, the very possibility of civilization itself, of "Prometheus, or the Promise of Provision" (297.L) depends on "Canine Venus sublimated to Aulidic Aphrodite" (299.L). While Shem continues to investigate the trauma of irreducible difference "down there on the batom," Shaun averts his eyes and gazes upward to a realm of pure mind in which his "apexojesus will be a point of order" (296.8–11). Shem recognizes the reality of the body—"Her trunk's not her brainbox" (299.18–19)—but Shaun recoils from his biological origin. In his self-serving *exegesis,* this "simpletop domefool" (299.15–16) pretends Anna Livia is a wholly spiritual being. Yet when he is unable to obscure completely her carnal nature and, consequently, his own worldliness, Shaun blames his mother for Man's fall into matter. He will henceforth identify exclusively with his father: "While that Other by the halp of his creactive mind offered to deleberate the mass from the booty of fight, our Same with the holp of the bounty of food sought to delubberate the mess from his corructive mund" (300.20–24). Shem's *creative reaction* to the discovery of *ALP*'s materiality is to seek, with her *help,* a means of *delivering* the *masses* from endless *strife.* Shaun, on the other hand,

42. As should by now be apparent, Joyce derives his farcically reductive rendering of human motivation from Freud's insistence that an erogenous zone of the body underlies every cultural manifestation. In his 1980 film *My Uncle in America,* Alain Resnais achieves a comparable effect when he employs the theories of the French behavioral psychologist Henri Laborit to establish an equivalence between human social conduct and the behavior of laboratory rats.

ingests the teachings of his father which so *corrupt* him that he *censors* the truth about Earwicker whose career now becomes "A PANOPTICAL PURVIEW OF POLITICAL PROGRESS" (272.R). Unlike Shem, for whom evolution is equally devolution, Shaun interprets history's futile succession of power struggles as a spiritual ascent toward perfection.

Joyce associates Shaun with a number of cultural figures in whom he detects an aspiration for permanent authority. Shaun can be compared, for instance, to Aristotle whose name in the language of the *Wake*—"Harrystotalies" (110.17)—suggests the deluded nature of the attempt to discover "A Place for Everything and Everything in Its Place" (306.17–18). Joyce also refers to several of his contemporaries. Thus, W. B. Yeats is mentioned because of his desire to restore in *A Vision*—"Gyre O, gyre O, gyrotundo!" (295.23–24)—the hierarchical cosmology of pre-Copernican Europe.[43] "Nightlessons" derides this nostalgia for the Great Chain of Being as "specious aristmystic" (293.18), or *specious aristocratic mysticism.* Similarly, Wyndham Lewis's *Time and Western Man,* which contains a denunciation of Bergsonian process philosophy, is parodically rendered "*Spice and Westend Woman*" (292.6), thereby implying a connection between Lewis's time-phobia and his notorious misogyny. Both stem from a reluctance to concede the priority of mutable matter. For Joyce, the desire to fix the meaning of the world and so transcend history emanates from "the beast of boredom, common sense" which pompously insists "in half a sylb, helf a solb, holf a salb onward" that "you must . . . in undivided reawlity draw the line somewhawre" (292.27–28, 31–32).

Shaun angrily draws that line precisely so as to banish Shem and his disquieting sexual curiosity beyond the pale: "You know, you were always one of the bright ones, since a foot made you an unmentionable, fakes!" (300.1–3). Shaun thus identifies Shem with *Oedipus.* But if Shem's interest in Anna Livia is an incestuous one, it is

43. Yeats puts it this way: "Much of the confusion of modern philosophy . . . comes from our renouncing the ancient hierarchy of beings from man up to the One"; quoted by Northrop Frye in his essay "The Rising of the Moon: A Study of *A Vision*"; see Denis Donoghue and J. R. Mulryne, eds., *An Honoured Guest: New Essays on W. B. Yeats* (London: Edward Arnold, 1965), p. 31.

Shaun who completes the Oedipal itinerary according to the conventional psychoanalytic account. That is, when the thunderer intrudes to assert his exclusive possession of Anna Livia—"ENTER THE COP AND HOW" (306.R)—threatening those "golden youths that wanted gelding" (20.30) with castration, Shaun abandons his quest for consummate satisfaction—"Abnegation is Adaptation" (306.L)—but only temporarily. He now takes his father's intimidating presence as a model to be emulated, hoping himself to become one day a "tidled boare" (297.31) powerful enough to subdue the world to his desire. He is therefore outraged when Shem mocks tradition and authority by forging the signatures of a succession of Irish cultural heroes—"This is Steal, this is Barke, this is Shwipt, this is Wiles, this is Pshaw, this is Doubbllinnbbay Yates" (303.5–8)—thereby demonstrating that personal identity for this "aboleshqvick" (302.18) is fluid, a process of stylistic mutation. Shaun responds violently to Shem's revelation of the self's potentially protean character: "In effect I could engage in an energument over you till you were republically royally toobally prussic blue in the shirt after" (305.12–15). He would punish Shem for his impious refusal to venerate the past by demanding the fetishistic obedience to tradition of an Irish *blueshirt* or fascist.[44]

In contrast to Shaun's promotion of national tradition, Shem is precisely a boundary crosser, a "Europasianised Afferyank" (191.4) who transgresses cultural difference and social hierarchy in order to bring an end to the "impovernment of the booble by the bauble for the bubble" (273.7). Shem hopes to abolish every form of patriarchal domination, including the linguistic: "He would wipe alley english spooker, multiphoniaksically spuking, off the face of the erse" (178.6–7). The putative author of *Finnegans Wake*, "Shem the Penman" (125.23) is at home in every language, which is equiv-

44. Although *Finnegans Wake* can on one level be understood as a pancultural "map of the souls' groupography" (476.33), Ireland provides a recurrent point of reference, the *setting*, so to speak, of this "nightynovel"'s (54.21) dream vision of human destiny. In *James Joyce and the Revolution of the Word* (London: Macmillan, 1978), Colin MacCabe makes the point that Shaun is finally provoked to violence by Shem's irreverent attitude toward the Irish national pantheon. Freud associates fetishism with nationalism in his essay "Fetishism," *SE*, 21:153–57.

alent to saying he is at home in no language. Unwilling to reside permanently within the confines of any system of meaning, he would subvert the law of the father by remaining always on the move from one interpretive standpoint to another. That is, he circulates among diverse cultural traditions, undermining the authority of each in turn. He is therefore excoriated throughout the *Wake* on account of the "jetsam of litterage" (292.16) in which he "scrabbled and scratched and scriobbled and skrevened nameless shamelessness about everybody ever he met" (182.13–14): "You and your gift of your gaft of your garbage abaht our Farvver!" (93.19–20). Shem is a "seeker of the nest of evil in the bosom of a good word" (189.28–29) and we are advised by Shaun, who would preserve his father's good reputation, to "Shun the Punman!" (93.13).

Shem defends himself with this rejoinder: "I'm only out for celebridging over the guilt of the gap in your hiscitendency" (305.8–9). By *celebrating* the guilt-ridden *hesitancy* in Shaun's speech pattern, Shem would *bridge over* or avoid his brother's futile effort to deny human being in the world: Shaun's *tendency* to conform like a model *citizen* to the example set by his father. Shem is precisely a "Shamman" (192.23) since he challenges the pretensions to spiritual authority upon which patriarchal rule depends. Thus, when he asks what might in other circumstances seem an innocuous question, "How do you spell 'hesitancy'?" (in Latin, "come si compita 'cunctatio'?"), Shem's query is really directed to the father's guilty recognition of his own desiring nature and consequently ensues in a scatological revelation: "COME SI COMPITA CUNCTITITITILATIO? CONKERY CUNK, THIGHTHIGHTTICKELLYTHIGH, LIGGERILAG, TITTERITOT, LEG IN A TEE, LUG IN A LAW, TWO AT A TIE, THREE ON A THRICKY TILL OHIO OHIO IOIOMISS" (305.R).

Shem is also a *shaman,* a role he assumes even as Shaun accords himself the patriarchal prerogative of violently suppressing dissidence. When Shaun finally strikes him on the jaw, Shem exclaims, "at the present momentum, potential as I am, I'm seeing rayingbogeys rings around me" (304.7–9). The *rainbow* is a traditional emblem of the shaman's visionary potency. It is, in fact, a bridge over which the shaman may pass in order to reach the "Other Side." Shem thus trespasses across the supposedly impassible divide separating spirit from matter. If Shaun is identified with Aristotle, Shem can be placed alongside Ovid, "The Voice of Nature in the Forest" (306.20–21). His purpose in writing the *Wake* is to return culture to nature, to resolve the antagonism of the two realms by showing how the spiritual is derived from the material. He is aptly described, therefore, as a "pelagiarist" (182.3): Shem's *plagiarism* of "thee faroots hof cullchaw" (303.20–21) is motivated by his *Pelagianism,* a heresy according to which free will can overcome "original sin," understood here as the deluded denial of humanity's intrinsic carnality.

This "meager suckling of gert stoan" (287.18–19) or *Gertrude Stein* does not, like Wyndham Lewis, shrink from the prospect of temporal process. Shem refuses to eternalize history by attributing to it a fixed pattern:

> The tasks above are as the flasks below, saith the emerald canticle of Hermes and all's loth and pleastir, are we told, on excellent inkbottle authority, solarsystemised, seriolcosmically, in a more and more almightily expanding universe under one, there is rhymeless reason to believe, original sun. (263.21–27)

By transforming an alchemical reverie about universal correspondence into an aside about the carnal motives ("the flasks below") of cultural activity ("the tasks above"), Shem ridicules the thunderer's rationalization of the nightmare of history. He would thus have Shaun *wake up* from the dream that all is *love* and *pleasure,* that harmony prevails under the thunderer's reign. Shem disputes Shaun's belief in a providential design by introducing the concept of a *serial* universe expanding *without rhyme or reason.* If history is cyclical, it merely repeats the thunderer's successive attempts to impose patriarchal hierarchy on a dynamic that relentlessly exceeds every effort to contain it. Shem concludes by blaming Earwicker for trying and praising him for failing to transcend the material world. "O felicitous culpability" (263.29), the fall is a fortunate one because history's archetypal architect never really controls the course of events. Hence, "GNOSIS OF PRECREATE DETERMINATION. AGNOSIS OF POSTCREATE DETERMINISM" (262.R). Even if the beginning of time can become an object of certain knowledge, history's end remains in doubt.

Shem's resistance to authority, his refusal to privilege any tradition extends to his own writing. According to the account of his career contained in the *Wake,* "He winged away on a wildgoup's chase across the kathartic ocean and made synthetic ink and sensitive paper for his own end out of his wit's waste" (185.5–8). Embarked on what may well turn out to be a *wild goose chase,* this "tragic jester" (171.15) hopes to effect a *cathartic* release from patriarchal rule by creating the *Wake* from his own "wit's waste"—a phrase which here has a literal significance—with a "highly prosy, crap in his hand, sorry!" Shem excretes "a no uncertain quantity of obscene matter" from "the bowels of his misery" (185.17–18, 29–30, 33). *Finnegans Wake* is written, that is, with "indelible ink" (185.26) made from urine and feces. By this "corrosive sublimation" (185.36) our "first till last alshemist" (185.34–35) accomplishes an *alchemical* transmutation in reverse, changing gold into lead: poetic vision has its basis in anal eroticism. Shem thus repudiates his own aspiration for authority by reminding his readers of the earliest form of idol worship.

❧

The present attempt to demonstrate an ideally speculative interpretive practice will now juxtapose a further tradition of thought to those already invoked in order to renew the possibilities of invention. Conceived according to an aesthetic of sublimity that values discontinuity and contrasting perspective rather than beautifully cohesive form, it will once more interrupt sequential exposition to begin again from a new point of departure. This Proof section has thus far advanced by the following stages: at the outset, a discussion of Joyce's festive overthrow of patriarchal authority in *Finnegans Wake* inspired it, as it were, to unleash its own will-to-invent. It therefore abruptly changed course to entertain a psychoanalytic theory concerning interpretive mobility. The ensuing reflection on the sublimation of desire then provoked a return to the *Wake* where Shem the Penman—"Pain the Shamman!"—was identified as the member of the Earwicker family most capable of such mobility. The foregoing discovery provides an additional opportunity for renewed invention. Proceeding on the basis of purely analogical continuity—just as it earlier moved back and forth between Joycean and psychoanalytic modes of discourse—this improvisational theater of ideas will now suppose that the phenomenon of shamanism represents an alternative to the patriarch's inflexible rule. A new interpretive excursion—this time employing the language of anthropology—will explore the suggestion that the figure of the shaman embodies the inventive ideal.

The Finnish national poem the *Kalevala* provides a convenient reference point for a discussion of the phenomenon of shamanism. The *Kalevala*'s chief personages are precisely shamans, and the events related in the poem express their characteristic aspirations and vision of the world. Yet when Elias Lönnrot first began collecting the songs he would later weave together to form the final 1849 version of the *Kalevala,* he understood his task as the recovery of the lost epic of Finland's heroic age. Lönnrot was concerned to justify Finnish longings for national identity and attempted, accordingly, to assemble a narrative that would recount, in terms of the career of a hero, the origins of the Finnish nation. This hero was Väinamöinen, for Lönnrot a genuine historical figure, the "First Father" of the Finns. Väinamöinen was to be for the *Kalevala* what Achilles, Siegfried, Beowulf, and Roland were for their respective epics. Lönnrot's own title for the unpublished *Proto-Kalevala* was, in fact, *The Collected Songs of Väinamöinen.* Lönnrot held that the original songs recounting the events of Väinamöinen's life had become corrupted by imperfect word-of-mouth transmission during the long centuries of Swedish domination. Consequently, he envisioned his mission as the faithful reconstitution of a lost "parent" text on the basis of which contemporary society might found itself anew. In the ancient heroic lays celebrating Väinamöinen's martial exploits, Lönnrot sought ancestral norms expressing the essence of the Finnish national character.[45]

But as he became increasingly well versed in the traditional mate-

45. See Felix J. Oinas, "The Finnish and Estonian Folk Epic," *Journal of Baltic Studies* 7, no. 1 (Spring 1976): 1–12; Taima Ranta, "Finnish Folklore," *Michigan Academician* 3, no. 3 (Winter 1971): 23–40; and William A. Wilson, "The *Kalevala* and Finnish Politics," *Indiana University Folklore Institute Journal* 12 (1975): 131–55. For the most useful English edition of the *Kalevala,* see Francis Peabody Magoun, Jr., trans., *The Kalevala or Poems of the Kaleva District,* compiled by Elias Lönnrot (Cambridge, Mass.: Harvard University Press, 1963).

rial of the Finnish peasants who were his principal informants, Lönnrot was to move away from pseudohistory and heroic legend and allow the tradition to speak for itself, so that the final version of the *Kalevala* embodies the *ethos* of the peasant society in which it originated. He thus gradually overcame his predisposition to regard his material as the product of a society in which a warrior aristocracy seeks fame for itself and its ancestors in epic song. In this sense, Lönnrot's growing competence as a singer of tales can be traced in his successive revisions of his material from the *Proto-Kalevala* through the *Old* to the *New Kalevala.*

A crucial difference between the poem's earlier and later versions involves the arrangement of events in the plot. At the end of the *Old Kalevala,* we witness Väinamöinen's defeat of Joukahainen and unhappy courtship of Aino. These are the penultimate episodes of the *Old Kalevala* and underscore Väinamöinen's putative centrality as the hero of the poem. In the *New Kalevala,* Lönnrot implicitly displaces Väinamöinen from the center of the poem by moving these episodes to the front of the *Kalevala,* where they are subordinated to the following narrative design: in the first half of the poem, the *Kalevala*'s three principal characters, Väinamöinen, Ilmarinen, and Lemminkäinen, each woo the maiden of North Farm. Ilmarinen eventually wins her hand in return for the Sampo, a magical device akin to the proverbial horn of plenty. Following the death of Ilmarinen's wife in the second half of the poem, Väinamöinen leads an expedition to recover the Sampo, in the course of which the Sampo is irretrievably lost. Ilmarinen's wedding to the maiden of North Farm thus becomes the central event of the *Kalevala.* The climax of narrative development in the first half of the poem, the wedding lays also motivate the action leading to the *Kalevala*'s final conclusion.

The wedding lays represent not merely the importance a peasant society attaches to exogamous unions. Ilmarinen's attempt to normalize relations between the Kaleva District and North Farm can also be interpreted from the standpoint of myth. Lönnrot himself describes the *Kalevala* as a work "future generations will possibly hold in as great esteem as the Germanic nations esteem the *Poetic Edda,* and Greeks and Romans esteem, if not precisely Homer, at least Hesiod."[46] The cycle of songs devoted to Lemminkäinen provides the most striking evidence for the hypothesis of the *Kalevala*'s mythological significance. In his song cycle, Lemminkäinen, wooing the maiden of North Farm, is assigned a number of difficult tasks in the performance of which he is killed and cut into pieces. His mother, hearing of his death, rushes to the scene and reassembles his body, thus restoring him to life. This sequence of events is a close replica of a shamanic initiation which, according to Mircea Eliade, "almost always includes one or more of the following themes: dismemberment of the body, followed by a renewal of the internal organs and viscera; ascent to the sky and dialogue with the gods or spirits; descent to the underworld and conversations with spirits and the souls of dead shamans."[47] The two other principal characters of the *Kalevala* also bear signs of their status as shamans. The orphic singer and enchanter Väinamöinen talks to birds, and particularly to the eagle, which is traditionally associated with the figure of the shaman. Similarly, Ilmarinen, the inventor of the Sampo, is identified as a smith, and for traditional societies the smith, like the shaman proper, is a figure endowed with mysterious abilities. At the furthest possible remove from warrior chieftains, then, Väinamöinen, Ilmarinen and Lemminkäinen are more properly thought of as sorcerers, magicians, seers who may pass from one plane of existence to another, from the underworld to this world to the paradise beyond.

The shaman's social role is thus intimately bound up with the process of myth. In an application of Claude Lévi-Strauss's structuralist myth analysis to the book of Genesis, Edmund Leach remarks on a persistent pattern in mythological thought:

> The myth first discriminates between gods and men and then becomes preoccupied with the relations and intermediaries which link men and gods together. . . . In every myth system we will find a persistent sequence of binary discriminations as between human/superhuman, mortal/immortal, male/female, le-

46. Quoted in Magoun's edition of the *Kalevala*, p. 352.

47. Mircea Eliade, *Shamanism: Archaic Techniques of Ecstasy* (New York: Pantheon, 1964), p. 34.

> gitimate/illegitimate, good/bad . . . followed by a mediation of the paired categories thus distinguished. "Mediation" (in this sense) is always achieved by introducing a third category which is "abnormal" or "anomalous" in terms of ordinary "rational" categories. Thus myths are full of fabulous monsters, incarnate gods, virgin mothers. This middle ground is abnormal, non-natural, holy. It is typically the focus of all taboo and ritual observance.[48]

The tribal magician or shaman occupies precisely this middle ground. As mythic mediator, it is the shaman who is able to travel from Earth to Heaven, from the merely natural to the supernatural. As Eliade puts it, "After the interruption of the *easy communications* that, at the dawn of time, existed between mankind and the gods, certain privileged beings (and first of all the shamans) preserved the power to actualize for their own persons, the connection with the upper regions." In a moment of ecstatic communion with the totality of existence, the shaman "abolishes the present human condition, and, for the time being, recovers the situation as it was at the beginning. Friendship with animals, knowledge of their language, transformation into an animal are so many signs that the shaman has re-established the 'paradisal' situation lost at the dawn of time."[49] One might say, from a psychoanalytic point of view, that shamanic ecstasy is a return to the mobile desiring fantasy of the "primary process."

According to this view, the unconscious never accepts the binding or structuring of desiring energy accomplished by the "secondary process" of the ego in order to enable intentional, project-oriented behavior. Rationality is always experienced, at least in part, as constraint, self-division, and loss of being. The shaman would thus be that special figure who is able "to transcend opposites, to abolish the polarity typical of the human condition, in order to

48. Edmund Leach, *"Genesis as Myth" and Other Essays* (London: Jonathan Cape, 1969), pp. 10–11.

49. Eliade, pp. 265 and 99. For Eliade, shamanism is the primordial form of mystical experience, as ancient perhaps as the paleolithic. The suggestion has been made that evidence for shamanic practices can be found in the caves at Lascaux; see, for instance, T. G. E. Powell, *Prehistoric Art* (New York: Praeger, 1966), pp. 62–63.

attain to ultimate reality."[50] For the shaman, the law of the excluded middle, the logical categories inaugurating this world and the next do not hold. Both Self and Other, this protean shapeshifter is not subject to the antithesis "human/nonhuman" and can therefore communicate with and even become whatever he or she beholds. In the figure of the shaman, the human community enjoys a vicarious recovery, in Breton's phrase, of the "omnipotence of dream."

The shaman's ecstatic affirmation of the here-and-now has, however, an unfortunate consequence. The phenomenal world is experienced as a reality so final that no action to alter the present is possible. As Robert Bellah suggests in his essay "Religious Evolution," the shaman is rather more the product than the producer of history. At the risk of misrepresenting a subtle exposition, Bellah's account of the evolution of consciousness can be summarized as a sequence of three stages: primitive, historic, and modern. Primitive religion corresponds to the moment of shamanic identification with immediate reality. During what Australian Aborigines call "the Dreaming," the gap separating subject and object, spirit and matter, disappears "when every when becomes now." According to Bellah, this paradisal experience of immanence "gives little leverage from which to change the world" because the savage mind "can only accept the world in its manifold givenness." The task facing historic religion is therefore to devise a strategy for resisting the world's intractable presence. The mind thus posits an eternal, immutable "beyond" as point of reference. That is, in historic religion, "devaluation of the empirical world and the empirical self highlights the conception of a responsible self, a core self, deeper than the flux of everyday experience facing a reality over against itself, a reality which has a consistency belied by the fluctuations of mere sensory impressions," a *transcendent* reality enabling the self to adopt a critical stance toward its mundane surroundings.[51]

But a new problem emerges: the possibility of transcendence depends on estrangement from the material world. Dogmatically attached to an unchanging idea, the self becomes alienated from its

50. Eliade, p. 486.
51. Robert Bellah, "Religious Evolution," in *Beyond Belief: Essays on Religion in a Post-Traditional World* (New York: Harper and Row, 1970), pp. 28, 29, 33.

environment with the result that effective action is once more precluded. The innovation brought about by modern religion consequently involves a reaffirmation of phenomenal reality. This return to the world is not, however, a restoration of "primitive monism," for the self no longer uncritically accedes to the present state of affairs. Nor can it be understood as a disguised repetition of the "hierarchic dualistic religious symbol system of the classical historic type" because paradise has been converted from a *spatially* distant and inaccessible place to an intention whose realization is *temporally* (or temporarily) deferred. Yet the modern phase in religious evolution is compounded equally of primitive and historic consciousness. It unfolds as a dynamic of immanence and transcendence. Thus, Bellah remarks, in the modern period "culture and personality themselves have come to be viewed as endlessly revisable."[52]

The *Kalevala* records the transition from primitive to historic religion. Lönnrot's poem also reveals why this passage from one form of consciousness to another fails to ensue in a productive interplay. The shamans of the *Kalevala* are flawed precursors of modernity. Mobile themselves, they nevertheless seek to immobilize the dynamic of immanence and transcendence so that presence and perfection will henceforth coincide. They thus remain ignorant of the historical basis of human being. Confirmation for this view depends on establishing the mythological significance of the Sampo. The Sampo has at least two meanings. First of all, it is a horn of plenty, emblem of the perfected condition of the other world. But the word itself means "pillar" or "post." In the traditional religion of the Ural-Altaic peoples to whom we owe the term "shaman," a pillar is the ritualistic symbol both for the fact of the various levels

52. Ibid., pp. 40, 44. In *The Mismeasure of Man* (New York: Norton, 1981), Stephen Jay Gould argues that natural selection has "worked for flexibility in human evolution": "wide behavioral ranges" have arisen "as consequences of the evolution and structural organization of the brain." Humans have evolved by "neoteny." That is, adult humans, who physically resemble the young of other primates, also retain juvenile behavioral characteristics such as curiosity and play: "Humans are learning animals"; pp. 331, 333. For Gould, the endless self-revision of modern consciousness can therefore be said to realize a deep-rooted biological imperative. Humanity thus lives up to its better nature.

of reality and for the possibility of the shaman's movement among them. The symbolic import of the pillar can be demonstrated in its convergence with the meanings of several other mythological images: "The tree along which the Altai shaman rises into Heaven," like the one that carries Ilmarinen to North Farm, "though furnished with divisions, is not really a post, but a leaf-crowned birch-tree. Thus, we find here an intermediate stage between the . . . world-pillars and the branched world-tree supposed to rise from the center of the earth."[53] In other words, in a cluster of related symbols, we have the idea of the world-tree, the *axis mundi,* the separation of reality into discrete levels, and the route of transit from one level to another. The *axis mundi,* the shaman's path of ascent to the other world, traditionally passes through the Pole Star, and would therefore be located in the vicinity of North Farm.

A cosmological scheme, a myth about the constitution of reality thus underlies the narrative pattern of the *Kalevala.* According to this myth, the separation of reality into discrete levels is the issue of human agency. North Farm is only rendered a paradise when Ilmarinen forges the Sampo in the hope of obtaining the maiden of North Farm as his bride. In this sense, the *Kalevala* can be read as a fable on the consequences of consciousness, understood as the ability to posit a world better than this one. The act of this positing is properly a fall, a fall into knowledge. The project subsequently pursued by each of the main characters of the *Kalevala* involves the attempt to establish a stable relationship of exchange (imaged in their efforts to win the maiden of North Farm) with the perfected other world. The *Kalevala* depicts, then, a magical project for the perfection of this world to be achieved by "marrying" it to the other one. It also suggests the futility of this effort, since to bring the two worlds together is not to use the one to perfect the other but only to destroy their difference, or the very possibility of conceiving of perfection. Paradise is always lost. And with the destruction of the Sampo, even the possibility of a shamanic transport to Heaven be-

53. Uno Holmberg, *Finno-Ugric, Siberian Mythology,* vol. 4 of *The Mythology of All Races,* ed. Louis Herbert Gray (New York: Cooper Square Publishers, 1964), p. 340.

comes precarious. The *Kalevala* thus ends on a pessimistic note: the departure of the failed magician Väinamöinen inaugurates a new epoch in which the fallen world of empirical existence is hopelessly distant from paradise, that better place now unattainably elsewhere.

Väinamöinen, Ilmarinen, and Lemminkäinen's doomed attempt to magically perfect the world has its basis in their fantasy that perfection might be achieved once-and-for-all, in their nostalgia, as it were, for a retrospectively perfected primordial state whose intrinsic contrariety has been forgotten. With reference to Euripides' *Bacchae,* one might say the shamans of the *Kalevala* fail to appreciate the following inescapable dilemma: if, on the one hand, Pentheus's hostility to instinctual reality can only ensue in his humiliation by the god, on the other, the Bacchants' Dionysiac rapture entails the abolition of consciousness, an incoherent frenzy of unfettered desire. Presence and perfection can never be made to coincide. The shaman must therefore worship both Aphrodite and Athena, Apollo and Dionysus. Lönnrot's aspiration for a heroic epic was not a misplaced one. Väinamöinen should have been a warrior chieftain. The shaman, who would sink back into the flux, must be complemented by the warrior, ready to struggle with the world.[54] Instead of seeking to identify the opposites, the shaman must consent to psychic division, must practice what Castaneda's Don Juan calls the "controlled folly" of a figure variously described as warrior and as sorcerer: "It's possible to insist, to properly insist, even though we know that what we're doing is useless. But we must know first that our acts are useless and yet we must act as if we didn't know it." For such a figure, life becomes an unending "exercise in strategy."[55]

❧

In "The Mime of Mick, Nick and the Maggies" (219–59), a chapter of *Finnegans Wake* for which the *Wake* itself supplies the title (219.18–19), Isobel becomes the object of Shem and Shaun's

54. To put it another way, the sense of visionary exultation with which Keats concludes *Hyperion*—"Knowledge enormous makes a God of me"—needs to be counterpointed by E. T. A. Hoffmann's ironic recognition of the shaman's limitations and vulnerability as revealed by the comically inept protagonist of *The Golden Pot.*

55. Carlos Castaneda, *A Separate Reality* (Harmondsworth, Middlesex: Penguin, 1973), pp. 82, 188.

sibling rivalry. The twins vie for the affection of their sister in a competition from which Shaun emerges as the victor. Isobel's companions, the rainbow girls—"R is Rubretta and A is Arancia, Y is for Yilla and N for GreeneriN. B is Boyblue with odalisque O while W waters the fleurettes of novembrance" (226.30–33)—dance round Shaun in "heliolatry" (237.1). The "Maggies" are *sun-worshippers* and therefore turn away from Shem, whose fatal shortcoming is that he is "not aebel to speel eelyotripes" (303.F1). A *heliotrope* herself, Isobel cannot resist inclining toward Shaun who, as her father's chosen successor, can be compared to the rising sun. She thus prefers "Mick" to "Nick," Saint Michael to Lucifer, the dragonslayer to the snakeworshipper. Like her young friends, she would rather believe that evolution is progressive—"Charley, you're my darwing! So sing they sequent the assent of man" (252.28–29)—that Darwin's *Descent of Man* is really an ascent toward perfection and not, as Shem the Penman covertly suggests, the inevitable decline and fall of a pompous *ass*. For Isobel, Shem's proclamation of the vanity of all things renders him ineffectual. The "ambitrickster" (423.6) is so derisive toward every authority he is incapable of taking positive action in support of any point of view. Consequently, she chooses Shaun, whose confidence in tradition encourages him to attempt a reformation of the world.

But though Shaun wins his sister's approval, he can do so only with the connivance, as it were, of Shem. Shaun is himself such a "filial fearer" (604.28) that he could not oppose his father's will, could not rescue Isobel from "earwigger's" (579.25) "ensectuous" (29.30) control without Shem, whose role throughout history has been that of the revolutionary propagandist who "coached rebelliumtending mikes" (287.29). Shem's farcical derision of the father liberates Shaun from the burden of the past, thereby enabling him to initiate a new cycle of tragic blunders. Shem returns consciousness to its material occasion so that Shaun, representing spiritual aspiration, might find a new point of departure, a new route toward an always distant ideal. Taken together, the fraternal contraries can be understood to comprise the twin poles of a single identity: precisely, Humphrey Chimpden Earwicker the "patriarchal shamanah" (75.14).

This necessary co-presence of Shem and Shaun does not, however, imply their harmonious cooperation: "Enten eller, either or. And? Nay, rather!" (281.26–29). Their relationship remains an endless struggle for dominance, "as by the providential warring . . . of dramdrinker against freethinker our social something bowls along bumpily, experiencing a jolting series of prearranged disappointments, down the long lane of . . . generations, more generations and still more generations" (107.30–35). In a passage referring to the "most ancient wisdom" of Bruno and Vico (287.20–28), Joyce describes Shem and Shaun as the rival banks of a great river—the River Liffey or stream of time surging through the "Eternal City" of Dublin—whose ceaseless flow prevents the peaceful unification of Earwicker's split personality.[56] Rivers have been an emblem of flux at least since Heraclitus. Anna Livia Plurabelle, "Bringer of Plurabilities" (104.2), incarnates the principle of flux as her husband embodies the will to permanence and stability. As representative of the force that animates the world, *Finnegans Wake* defers to "Annah the Allmaziful" (104.1) at its outset: "riverrun" (3.1), concatenating a noun and a verb, is suggestive of a processual dynamic that never settles upon a definitive stage of realization. Hence, "madernacerution" (107.19) or *mother-nation-revolution* locates Anna's agency in human affairs in the endless transformations of society that are analogous to the language of the *Wake* in which coherence is only temporary among a jostling hubbub of cultural traditions—

> Every person, place and thing in the chaosmos of Alle anyway connected with the gobblydumped turkery was moving and changing every part of the time: the travelling inkhorn (possibly pot), the hare and turtle pen and paper, the continuously more and less intermisunderstanding minds of the anticollaborators, the as time went on as it will variously inflected, differently pronounced, otherwise spelled, changeable meaning vocable scriptsigns. (118.21–28)

Anna subverts the law of her husband, who would master the world by imposing upon it a uniform, unambiguous system of meaning, by providing for an ever-multiplying array of actors on the stage of

56. The passage is given in Latin; for a translation, see McHugh.

history each of whom arrives with an alternative point of view. "Gruff Gunne may blow, Gam Gonna flow" (271.17–18): even as Earwicker loudly insists on his preeminence, Anna steadily erodes the foundations of patriarchal rule. As "laotsey taotsey" (242.25–26) puts it: "Under heaven nothing is more soft and yielding than water. Yet for attacking the solid and strong, nothing is better."[57]

Already in her youth, she is well versed in "all the runes of the gamest game ever": "This isabella I'm on knows the ruelles of the rut and she don't fear andy mandy" (279.F1). The reality of sexual politics elicits from her no sentimental rationalization: "One must sell it to some one, the sacred name of love" (268.F1). If civilization begins for Earwicker when "shoo, his flutterby, / Was netted and named" (262.13–14), when women were defined as Man's subservient complement, Isobel retorts these "gelded ewes" will nevertheless go "jilting about" (279.F1). They will betray every patriarchal authority that seeks to render them powerless: "She'll confess it by her figure and she'll deny it to your face" (271.14–15). Anna thus encourages Shem to denounce the narcissistic and incestuous cravings that underlie Earwicker's pretensions to spiritual authority. At the same time, however, she urges Shaun to emulate his father. Like Earwicker, he should involve himself "in somes incontigruity coumplegs of heoponhurrish marrage from whose I most sublumbunate. A polog, my engl!" (607.20–22). *Apologizing* to his *angel* for the *incongruous* carnality he would sooner consider a *mirage,* he resolves to employ his desire in the service of the *sublimated* activities of civilized life. Hence, "obelisk rises when odalisks fall" (335.33) scatologically posits Anna Livia as inspiration for the "eternal chimerahunter" 's (107.14) *monumental* delusions. In this connection the rainbow associated with Anna can be seen to have a double significance: Shem crosses it one way to reach the realm of the senses while Shaun, traveling in the opposite direction, pursues a purely spiritual ideal.

ALP is able, then, to play the muse to all the men in her life: "Mimosa Multimimetica, the maymeaminning of maimoomeining! Elpis, thou fountain of the greeces, all shall speer theeward, from kongen in his canteenhus to knivers hind the knoll" (267.2–6). A

57. Lao-tze, chapter 78.

sensitive plant responsive to every stimulus, she is the sexual object of both the *king in his counting house* and the *knaves* who plot his overthrow. The real meaning of Ogden and Richard's *Meaning of Meaning* is the *stuttering love* of every man for whom Isobel is "our shibboleth" (267.21), the *stream in flood* that is also our *password* or *slogan*—

> All the world's in want and is writing a letters. A letter from a person to a place about a thing. And all the world's on wish to be carrying a letters. A letters to a king about a treasure from a cat. When men want to write a letters. Ten men, ton men, pen men, pun men, wont to rise a ladder. And den men, dun men, fen men, fun men, hen men, hun men wend to raze a leader. (278.13–21)

The cyclical raising up and overthrow of the sacred king ("raze a leader"), the attempt to ascend into the sky to become pure spirit ("rise a ladder"), and the recurrent effort to discover the truth about reality ("a letters") are all inspired by "a treasure from a cat." But if Anna Livia—"the pussy in the corner" (278.6)—provides their inspiration, none of these projects for transcending the vagaries of history can win her unqualified approval: they are to "be slipped on, to be slept by, to be conned to, to be kept up. And when you're done push the chain" (278.F5). Neither Shem the realist nor Shaun the idealist has her undying love. Though each may have his moment of glory, neither can permanently remain in the limelight: "Rockaby, babel, flatten a wall" (278.L). The *Tower of Babel,* emblem of the desire for transcendence, is built by Humpty Dumpty, dethroned when he falls from his *wall.* History oscillates between (Shaun's) veneration and (Shem's) derision of cultural tradition. To put it another way, even as she *rouses* her husband from dogmatic slumber, Anna Livia *arouses* him with new delusions of grandeur.[58]

Her relationship with Earwicker is thus thoroughly "ambiviolent" (518.2): "From gramma's grammar," Isobel learns there is "comfortism in the knowledge that often hate on first hearing comes

58. According to Joseph Campbell and Henry Morton Robinson, Anna's role in the *Wake* is first "to shatter HCE as the container of fixed energy" and then "to gather up the remains . . . and consign them to a fresh start"; see *A Skeleton Key to Finnegans Wake* (New York: Harcourt, Brace, 1944), p. 11.

of love by second sight" (268.17, 29–269.2). Love and hate are intermingled from the beginning. On the one hand, she must beware "how in that hist subtaile of schlangder lies liaison to tease oreilles" (270.15–16). Earwicker's *subtle history of slander* ("Singalingalying. Storiella as she is syung" [267.7–8]: the story about Anna Livia as "a reel of funnish ficts" [288.8–9]) cannot conceal the fact that the *snake* is precisely his *tail* ("Funny spot to have a fingey!" [144.35]). Isobel need not, therefore, accept responsibility for Man's fall into matter. But on the other hand, she unashamedly exults in her own carnality. Dreaming of "the glorifires of being presainted maid to majesty" (304.21–22), she recalls with pride her participation in pagan fertility rites: "Wasn't it just divining that dog of a dag in Skokholme as I sat astrid uppum their Drewitt's altar, as cooledas as culcumbre, slapping my straights till the sloping ruins, postillion, postallion, a swinge a swank, with you offering me clouts of illscents and them horners stagstruck on the leasward!" (279.F1). Thus moving from one contrary point of view to another, Isobel will have her "little sintalks in the dunk of subjunctions, dual in duel and prude with pruriel" (269.3–4). That is, she must come to terms with both Shaun and Shem, the prudish dueling with the prurient, spirit with matter in an irresolvable conflict. Her mother enjoins her to "mind your genderous towards his reflexives" (268.25): even as she welcomes or is *generous* toward Earwicker's desiring interest (for him, an automatic *reflex*) she should be *mindful* or suspicious of the consequences for her *gender* of his *reflexive* fantasy of invulnerable self-enclosure. As Anna Livia says, "Learn from that ancient tongue to be middle old modern to the minute" (270.17–18), to find a middle way between old and modern, tradition and the present, to remain fluid and unattached while yet recognizing the need for positive commitment.

Isobel must be unafraid, then, of self-contradiction. Who are Shem and Shaun, matter and spirit, and every antithetical pair "the living sons or daughters of?" Answer: "Night now!"; all spring from "the rivering waters of, hitherandthithering waters of. Night!" (216.3, 5–6). The everyday world of clear and distinct entities, the known universe of hierarchy and degree is a misleading simplification of the nightworld's endless variability, the nightworld whose tutelary deity is precisely Anna Livia Plurabelle. In *Finnegans Wake,*

Humphrey Chimpden Earwicker is "bewilderblissed" by her "night effluvia" (107.16, 17); he is *bewildered* (or made forgetful of his daytime identity) and *blissed* by the streaming productions ("a streamsbecoming" [597.8]) of that unforseen spontaneity which takes every would-be authority by surprise. The eternally frustrated male self, rigidly adhering to a partial truth—"man made static"—thus encounters an orgasmic flood of new possibilities—"woman formed mobile" (309.21–22). In other words, the twins make the following startling discovery in the course of their geometry lesson: "Mother of us all! . . . If that aint just the beatenest lay I ever see! And a superpbosition! Quoint a coincidence! O.K. *Omnius Kollidimus*" (299.3, 6–9). Everybody collides—"All the charictures in the drame" (302.31–32)—the whole panoply of interpretive standpoints are equally present in Anna Livia, the "mother of invention." "A Tullagrove pole is . . . revisible by nighttim, may be involted into the zeroic couplet" (284.5, 8–9): the *phallus,* emblem of Earwicker's dogmatic rigidity (and therefore *risible* for the duration of this "nightynovel") may be *reversed* or *inverted* into the "zeroic couplet," a reference to Shem and Shaun's geometric rendering of Anna Livia's sexual anatomy. By nighttime, Humphrey Chimpden Earwicker *is* Ann Livia Plurabelle.

In effect, the monologue of the primal father, and the dialogue of the parricidal twin sons, are dissolved in the polylogue, in Coleridge's phrase, of "Alph, the sacred river," precisely an *alphabet* encompassing the entire array of cultural traditions: "Every letter is a godsend, ardent Ares, brusque Boreas and glib Ganymede like zealous Zeus, the O'Meghisthest of all" (269.16–19). Anna Livia thus makes possible a form of consciousness in which the self's putative continuity becomes a succession of discontinuous instances of thought; a theatricalized self which, Leo Bersani says, "would, ideally, resist all interpretive efforts," every effort to decipher an unchanging human presence behind this presently talking head.[59] In Joyce's words, "In this scherzarade of one's thousand one nightinesses that sword of certainty that would indentifide the body never falls" (51.4–6); the ever-proliferating names circulate around the

59. Bersani, *A Future for Astyanax,* p. 238.

body without penetrating to a psychological core, without resolving in a static unity the multiple "identities in the writer complexus (for if the hand was one, the minds of active and agitated were more than so)" (114.33–35). Instead of the conventional itinerary of beginning-middle-and-end, Anna Livia enables that "disinterested play of thought" to which Breton and the surrealists aspired whose unfettered buoyancy is never permanently fixed either in a retrospective or a prospective order.[60] Anna Livia, "Brook of Life" (264.6), a marginal "outside" figure (outside the law of the father) has really always been the "inside." She is, in fact, the center which is no center because it stands for nothing fixed and definite but should be understood rather as a possibility of movement.

But as a possibility of movement among every position in the family constellation, our "eternal geomater"'s (296.31–297.1) structural identity fails to provide a model, a point of reference or *topos* in terms of which mental evolution might orient itself. If Earwicker is a historical dynamic kept in motion by Anna Livia, at any given moment Anna Livia is a historically contingent version of her husband longing for immortality: "All me life I have been lived among them" (627.16). Only as HCE—"Here Comes Everybody" (32.18–19)—can "she" take "his" place. To the extent that Anna Livia can be defined as the spontaneity of instinctual life in opposi-

60. The analogy made at the beginning of this Proof section of *Kaironomia* to the work of composer Philip Glass, who repeatedly ratifies a brief contour of thought in order to call attention to the listener's *present*, should therefore be supplemented with reference to a composer like John Cale whose expressionist theatricality ensues in a *history* of stances toward the world. That is, Anna Livia always inhabits the here-and-now because she is always about to move elsewhere. The same point could be made in terms of the classics of the modern canon by juxtaposing Gertrude Stein with Picasso or Stravinsky. Thus, while Stein characteristically offers an experience of radical sublimity—"Elephant beaten with candy and little pops and chews all bolts and reckless reckless rats, this is this"—Picasso, in *Les Demoiselles d'Avignon*, allows "that obscure object of desire" to remain inscrutably other by drawing upon culturally remote conceptions of human being in a work deliberately left unfinished, unresolved, perpetually in progress. Stravinsky's self-revising, restlessly inventive sensibility has recently been thematized by Pierre Boulez in a series of recordings emphasizing precisely the stylistic mutations in Stravinsky's career; see, for instance: Igor Stravinsky, *Chamber Works* (1918–1962, including *Ebony Concerto* and *Dumbarton Oaks*), cond. Pierre Boulez, Ensemble Intercontemporain, Deutsche Grammaphon, 2531–378, 1982.

tion to the self's dogmatic inflexibility, one might remark with Freud: "An instinct can never become an object of consciousness—only the idea that represents the instinct can."[61] Anna's ability to initiate change will not save us, then, from folly and delusion. She appears as an endless procession of Kuhnian paradigms, of limited perspectives each pretending to tell the whole truth. The equation can therefore be reversed: Anna Livia Plurabelle *is* Humphrey Chimpden Earwicker.

We are thus confronted with the "Co-Education of Animus and Anima" (307.3–4) in "Tiresias" (307.L), the visionary dreamer of *Finnegans Wake*. At the beginning of time, "Hoddum and Heave, our monsterbilker, balked his bawd of parodies" (296.6–7). Earwicker *cheated* his *bird of paradise* by projecting part of his own nature onto his partner: "Hoddum and Heave" is *Adam and Eve* in one person. The Rabelaisian giant Finn is androgynous, a commingling of male and female, spirit and matter, ego and id. The combination of Earwicker's ever-burning desire with Anna Livia's instinctual fluidity produces firewater or whiskey—"*Usqueadbaugham!*" (24.14)—in Gaelic, the water of life: "At Tam Fanagan's weak yat his still's going strang" (276.21–22), distilling great quantities of "Irish frisky" (470.33) so that "rivers burst out like weeming racesround joydrinks for the fewnrally, where every feaster's a foster's other, fiannians all" (277.3–6). As Ovid comments in the *Metamorphoses:* "Fire may fight with water, / But heat and moisture generate all things, / Their discord being productive."[62]

Beckett describes this as the *Wake*'s "purgatorial" aspect: "a flood of movement and vitality released by the conjunction" of antithetical elements, a purgatory in which "movement is non-directional—or multidirectional, and a step forward is, by definition, a step back."[63] That is, in the "nightynovel" of *Finnegans Wake* our primordial parents can exchange the features that distinguish them by daytime. Anna Livia is thus an "allaphbed" (18.18), a rubbish heap of dead cultural traditions covering the "brontoichthyan form outlined aslumbered" (7.20–21)—"THE MONGREL UNDER THE

61. Freud, "The Unconscious," *SE*, 14:177.

62. Ovid, *Metamorphoses*, trans, Rolfe Humphries (Bloomington: Indiana University Press, 1955), p. 16.

63. Beckett, "Dante . . . Bruno. Vico . . . Joyce" (p. 22).

DUNGMOUND" (276.R)—of her *thunder-fish* husband, that *dinosaur* only temporarily extinct. Anna Livia, the River Liffey, is also, then, an alp or mountain like Earwicker himself: "The marriage of Montan wetting his moll we know" (260.16–17). In other words, when we render her finite by containing her "plurability" within a name: she is her husband. Similarly, when she rejoins the sea—her "cold mad feary father" (628.2)—at the end of the *Wake,* we are reminded that history evolves according to the mutations of HCE: he is his wife. Each can become the other: neither identity is permanent, neither position fixed.

The meaning of *Finnegans Wake* is in a continual state, as Philippe Sollers has claimed, "of skidding, of lapsus" ("lapse not leashed" [63.24]).[64] We are unable to affirm one side of any opposition over another: "Heil, heptarched span of peace! Live, league of lex, nex and the mores!" (273.4–5). The "heptarched span" of Anna Livia's *rainbow,* which inspires Shem to write the *Wake,* can never displace the *morality, legal code,* and *violent deeds* of Humphrey Chimpden Earwicker, Shaun's idol. In this connection, Geoffrey Hartman maintains we cannot reside either "within a maternal or a paternal sphere. The former would result in a stilling that is absolute, the latter in a fixity that is absolute."[65] On the one hand, Anna Livia would flood the world with desire, dissolving every distinction, every form and structure, and consequently subside into entropic immobility. On the other, Earwicker's inhibited rigidity, his refusal to be *moved* by his experience must finally explode in violence. Our attitude toward Earwicker and his wife must therefore remain ambivalent. According to Stephen Heath in an essay titled "Ambiviolences," Joyce's bottomless irony "operates a dislocation of sense without resolution."[66] Or to borrow a phrase from Roland Barthes, *Finnegans Wake* speaks "the locus of meaning but . . . does not name it."[67]

64. Philippe Sollers, "Joyce & Co.," trans. Stephen Heath, *Triquarterly* 38 (1977): 112.

65. Geoffrey Hartman, *Saving the Text: Literature/Derrida/Philosophy* (Baltimore: Johns Hopkins University Press, 1981), p. 152.

66. Stephen Heath, "Ambiviolences: Notes pour la lecture de Joyce," *Tel Quel* 50 (1972): 27 (my translation).

67. Roland Barthes, *Critical Essays,* trans. Richard Howard (Evanston, Ill.: Northwestern University Press, 1972), p. 219.

Neither Humphrey Chimpden Earwicker nor Anna Livia Plurabelle nor Shem nor Shaun nor Isobel can ever achieve a commanding perspective on the circulation of meanings in *Finnegans Wake* because all are caught up in it. Programmatically undermining every point of view, the *Wake* would thus affirm the need for a speculative practice of beginning again and again, an improvisational activity of invention willing to risk radical departures from canonical tradition. This Proof section of *Kaironomia* has been inspired by a similar motive. It has therefore *mimed* the *Wake*'s complicated self-subverting structure by deploying it in the form of an interpretive drama. Successive passages commenting on the *Wake* have been narrated, as it were, by the members of the Earwicker family themselves, who deride one another for their fallibilities and shortcomings as they struggle for dominance. First, the father was repudiated by his parricidal twin sons who eagerly sought to succeed him; then, the sons were dismissed in favor of mother and daughter, both impartially dubious of every patriarchal scion; finally, mother and daughter, whose instinctive skepticism left them unable to act in a world of half-truths, were compelled to give way in turn. Shem and Shaun displaced Earwicker from the center of significance only to be displaced themselves by the equally fleeting centrality of Isobel and Anna. The stance toward experience associated with each member of the Earwicker family was superseded, then, in a self-subverting interpretive movement verging on limitless displacement.

This palpable proof of the provisional character of every standpoint should not, however, give rise to a fantasy of invulnerability in which speculative thought involves itself in contingent situations without ever becoming attached to a one-sided point of view. The mind must in fact emphatically be warned against the deluded belief that an awareness of universal delusion might somehow save it from folly; a belief, finally, in the possibility of being *in* the world and *above* it at the same time. One last turn of the screw remains to be accomplished: the present interpretive drama, which began with the festive overthrow of an archetypal Lord of Misrule, will now close with the restoration of this monarch to his throne. Silenced only for the moment by his rebellious offspring and spouse, the primordial thunderer will once more loudly proclaim his authority and preeminence.

Thus, even as the *Wake* upsets the hegemony of patriarchal rule, of mystified veneration for the past, it concedes the inevitability of Earwicker's homecoming: "We drames our dreams tell Bappy returns. And Sein annews" (277.17–18).[68] The *Wake* promises no permanent protection against dogma and cant, only a temporary respite from endless strife. It is, in fact, a ritual—"Strangely cult for this ceasing of the yore" (279.2–3)—in which, as Bellah says, by "acting out the mistakes and sufferings of the paradigmatic mythical hero, the new initiates come to terms symbolically with . . . the 'immemorial misdirection' of human life."[69] *Finnegans Wake* is a rite of spring—"Pingpong! There's the Belle for Sexaloitez!" (213.18–19) or Sechseläuten, a Swiss spring festival—both *tragic,* since it recounts the death of a hero, the dismemberment of a god, and *farcical,* since this Pulcinella figure—"pleased as Punch" and "hugly Judsys" (620.23, 26)—an eternally renascent, indestructible satyr will surely rise again: "lovesoftfun at Finnegan's Wake" (607.16).

If the dream-vision of *Finnegans Wake* offers a privileged moment free from delusive attachment, waking consciousness—"Into the wikeawades warld from sleep we are passing" (608.34)—remains implicated in error. The sublimation of desire that produces the world we *know* does so by obscuring reality. The wish to achieve an ideally mobile form of interpretive invention must struggle ceaselessly with the human propensity for self-delusion, destined to prevail once more when the *Wake* ends. Humphrey Chimpden Earwicker, that "manyfeast munificent more mob than man" (261.21–22), is precisely everyone's everyday self: "The hundering blundering dunderfunder of plundersundered manhood; behold, he returns; renascenent; fincarnate" (596.2–5). We wake from *Finnegans Wake* only to fall again into dogmatic slumber. Assimilating the

68. In this connection, one might remark that Earwicker has in fact never really left: acting through Joyce, he is himself the author of *Finnegans Wake.* Joyce's mania for exhaustiveness, the multitude of cultural figures who crowd the pages of the *Wake*—"bark," "beethoken" and "sweetmoztheart"; "Eyeinstye" and "spinosis"; "hakusay" and "remembrandts"; "Boccuccia," "Chacer" and "Shakhisbeard"; not to forget Popeye the Sailorman ("I yam as I yam")—all testify to a desire to write a definitive text embracing every previous tradition within its own magisterial perspective. For an extensive list, see Adaline Glasheen, *A Third Census of Finnegans Wake* (Berkeley and Los Angeles: University of California Press, 1977).

69. Bellah, "Religious Evolution," in *Beyond Belief,* p. 28.

sublime to the beautiful, direct experience of the *Wake* is replaced by a recollection. The farceur's affective freedom thus gives way to tragic identification with a partial truth. Hence, Joyce's concluding joke: on the one hand, the last word of *Finnegans Wake*—"the" (628.16)—points to *the* present, to the shock of the new that can never be put into words; but on the other, it announces a recurrence of *the* past, of personal tradition and quotidian routine. Thought can only transpire as a dynamic of "*Oropos Roxy and Pantharhea*" (513.21–22), of *Oedipus Rex and Everything Flows,* an interminable dynamic of fiction and truth, self and other, *nomos* and *kairos.*

PART FOUR

An Image of Change

Truthful words are not beautiful.
Beautiful words are not truthful.

—Lao-tze, *Tao Te Ching*

It is much easier to go along the sides, where the outer edge serves as a limit and a guide, than by the middle way, wide and open.

—Montaigne, "Of Experience"

In a series of drawings sketched toward the end of his life, Leonardo da Vinci imagined the inundation of the globe by a great Deluge, a flood cascading down from the heavens with such terrifying force, such irresistible momentum that neither civilization nor the natural landscape itself can escape being swept away. The drawings present an apocalyptic vision of flux and universal dissolution in which every form and structure, every distinction and degree is suddenly absorbed into a primordial chaos of swirling movement. For Leonardo, the Deluge transpires as a climactic explosion of pent-up energy, an abrupt and catastrophic acceleration of the principle of mutability propelling historical process: every conceivable change happens here-and-now. As whirlpools and convulsive surges of water consume the earth, all the dynamism inherent in nature is released to spend itself in a moment of frenzied metamorphosis.

The possibility of so fantastic and appalling a catastrophe recurs more than once in Leonardo's notebooks. Leonardo can frequently be observed, Kenneth Clark says, wholly engrossed in elaborate descriptions of natural calamity, looking forward to the end of the world with strangely eager anticipation as he recounts how frightened animals will cluster together on the last hilltops not yet submerged, how the flood waters will be strewn with the floating wreckage of civilization and with the bodies of the drowned, how surviving humans will struggle vainly to improvise life rafts amidst the buffeting of the waves, how the fury of howling tempests will lay waste to cities, towns, and every habitation. At one point, he concludes a catalogue of pictorial subjects with the following visions an artist might summon into being—

> and then rivers falling from high mountains with the force of great floods, ruins which drive down with them uprooted plants mixed with rocks, earth, and foam and wash away to its ruin all that comes in their path; and then the stormy sea, striving and

> wrestling with the winds which fight against it, raising itself up in superb waves which fall in ruins as the wind strikes at their roots.

Clark remarks in his commentary on this passage that Leonardo is here "carried away by his true feelings. . . . Instead of compositions of figures, classical and religious legends, beautiful faces and draperies, all the subjects which pleased the patrons and artists of his time, he describes this combat of the elements, a subject for Turner, in the language of Herman Melville."[1] Leonardo was fascinated, in other words, by the prospect of violent, cataclysmic change. As he put it himself, he could conceive of no spectacle more compelling than the "irremediable and destructive terrors . . . caused by rivers in flood." Why, then, this obsessive preoccupation with apocalyptic flux?

According to Jean Laplanche, the chaotically fluid dynamism of Leonardo's Deluge drawings evokes "a level of *pure instinct,* of pure force anterior (if that is imaginable) to every symbolization."[2] Not merely a depiction of catastrophic change in nature, this flux is also internal. The Deluge drawings can be understood to represent an unconscious sea of instinctual yearnings, relentlessly mobile and metamorphic, surging forth to overwhelm the entire universe. In his successive depictions of the end of the world, Leonardo da Vinci, magisterial *bricoleur* and inventor without peer, provides a portrait, or rather, a series of portraits of his own inventive will at that fateful turning when the will-to-invent need struggle no more because it has succeeded at last in exhausting the possibilities of invention: every occasion, every point of departure has quite literally been consumed. Leonardo dreams, that is, of swallowing up the totality of existence in the swirling vortex of his own voraciously appetitive desiring imagination. To put it another way, in the Deluge drawings Leonardo entertains a urethral fantasy of finally extinguishing the flames of desire in a definitive spending of instinctual energy; precisely, a

1. Kenneth Clark, *Leonardo da Vinci: An Account of His Development as an Artist* (Harmondsworth, Middlesex: Penguin, 1967), p. 81. The passage from Leonardo's notebook is translated by Clark, p. 80.

2. Laplanche, *Problématiques III,* p. 232.

flood of desiring impulses. Leonardo's fabled curiosity about all things is thus converted into a nihilistic urge to engulf the world, a climactic confrontation with experience abolishing time and the duality of self and other in an eternal Now of sublimely blissful Oneness. The Deluge drawings reveal his impatience with the interminable nature of invention, his wish to achieve a final form. The stream of Leonardo's consciousness, protean and shapeshifting, able to accommodate itself to any circumstance, to adapt to any situation, has nevertheless a suicidal motive: it seeks a conclusive moment in which every object of desire, all the possible meanings of existence will be present in a lasting simultaneity, a moment in which desire's agitated mobility will forever be stilled.

Yet the fluid motion of water was not originally for Leonardo an emblem of brutal, world-destroying flux. To the contrary, Leonardo regarded the rippling currents of swiftly flowing streams as paradigmatic of every graceful movement. As Kenneth Clark observes, Leonardo's eventually scientific interest in hydrodynamics arose from his wish to render movement as "the visible expression of grace." And for the Renaissance, Clark says, grace "implied a series of smooth transitions," a harmonious reconciliation between process and structure as revealed, for instance, by the shell of a chambered nautilus which in cross-section displays the history of the creature's growth as a spiraling procession of forms, precisely, as shaped vitality.[3]

Castiglione's *Book of the Courtier* provides a subtly nuanced exposition of this Renaissance conception of grace. At the beginning of *The Courtier,* the assembled members of the Court of Urbino decide to play a game with the object of defining the attributes a courtier must acquire in order to become a consummately graceful human being. Since the participants in this contest are themselves courtiers, their willingness to adopt a playful stance toward the question of their own identity suggests that life for them is ideally lived as a form of play. And playfulness does indeed emerge as an important component of graceful conduct when Castiglione's friend Count Ludovico da Canossa argues that a certain quality of effortless ease is particularly desirable in a courtier.

3. Clark, *Leonardo da Vinci,* p. 14.

> I have found quite a universal rule which in this matter seems to me valid above all others, and in all human affairs whether in word or deed: and that is . . . to practice in all things a certain *sprezzatura,* so as to conceal all art and make whatever is done or said appear to be without effort and almost without any thought about it. . . . Therefore we may call that art true art which does not seem to be art. . . . And I remember having read of certain most excellent orators in ancient times who, among the other things they did, tried to make everyone believe that they had no knowledge whatever of letters; and, dissembling their knowledge, they made their orations appear to be composed in the simplest manner and according to the dictates of nature and truth rather than of effort and art; which fact, had it been known, would have inspired in the minds of the people the fear that they could be duped by it.[4]

To display the nonchalance and unforced spontaneity of *sprezzatura,* a courtier must be responsive to his immediate situation, must possess that negative capability Arthur Symons attributed to Mallarmé, "an easily moved alertness of mental attitude." Only by acting on the spur of the moment—"without effort and almost without any thought about it"—can the courtier infuse his conduct with the effortless ease of a performance that is truly the issue of its time. He must therefore avoid whatever is premeditated, customary, or routine. Daniel Javitch remarks that the perfect courtier would continually test his "ideals against the claims of experience," and would always be able to make the most of the fortune of the moment thanks to his "flexibility in demeanor and point of view," like the assembled company at Urbino themselves, who demonstrate "in their style and in their temperament . . . a marked intolerance and distrust of absolutes or even settled convictions." In other words, if he would achieve an improvisational form of selfhood identical with its present performance, the courtier must assume an ironic stance toward the urgencies of personal desire. The situational, context-oriented consciousness of the ideal courtier would therefore exhibit, in addition to *sprezzatura,* a further attribute, what Castiglione calls "*mediocrità,* that equilibrium whereby a predominant trait deter-

4. Castiglione, *The Book of the Courtier,* trans. Charles S. Singleton (New York: Anchor, 1959), pp. 43–44.

mined by personality, office, or state of mind is tempered by its opposite." For the assembled members of the Court of Urbino, the courtier's conduct "is deemed most becoming when he disguises a particular intention, disposition, or ability by cultivating an appearance of its contrary," a feat of deliberate self-contradiction that preserves the courtier from one-sidedness, or dogmatic attachment to a relative point of view.[5] *Mediocrità* and *sprezzatura,* skepticism, and a readiness to improvise are equally essential, then, if the courtier is to enter the charmed circle of earthly grace.

The courtly *ethos* promotes irony, playfulness, and occasionality as a way of being in the world. But this ironic, playful, and occasional mode of comportment, so disarming because so obviously without ulterior motive, in actuality does serve a hidden purpose. While apparently practicing a sophistic-Gorgianic form of oratory as decorative display, the ideal courtier, Javitch argues, secretly affirms the Isocratean-Ciceronian view that eloquence entails a polemical thrust. Grace is a rhetorical strategy by means of which the self seeks to impose its will upon the world. The courtier's effortless ease, like the impromptu oratory of antiquity, depends on the calculated dissembling of previously acquired expertise. However much his playful and ironic manner may destabilize personal tradition, the courtier nevertheless intends to transport his audience beyond a customary standpoint and thus to destabilize collective tradition. According to Ottaviano Fregoso, perhaps the most outspoken moralist among Castiglione's courtiers, the perfect courtier would entice and beguile his prince with a salutory deception, a deception not merely pleasing but one that will instruct and even move courtly society to take some new course of action. Yet the deception upon which the appearance of effortless ease depends is never wholly concealed. Javitch notes, "*The Courtier* never advocates sheer deceit. If the courtier is encouraged to be deceptive, his dissimulation must never be so total that his authentic intentions fail to be recognized."[6] In fact, the courtier is precisely ironic about his own intentions, which become the occasion for a playful display of improvisa-

5. Daniel Javitch, *Poetry and Courtliness in Renaissance England* (Princeton: Princeton University Press, 1978), pp. 87, 86, 32, 56, 36.

6. Ibid., p. 57.

tional skill. From the very beginning, then, the illusory nature of the courtier's freedom from constraint is openly acknowledged. And by proclaiming his inability to transcend private obsession, the courtier achieves, paradoxically, a measure of transcendence. That is, when he in effect encourages dissent from his present point of view by adopting an ironic attitude toward himself, the courtier renews his commitment to dialogue as an ideally gracious form of social exchange.

The "motives of eloquence," in Richard Lanham's phrase, are thus complex. The courtier is caught in a double bind. His ceaseless oscillation between speculative fancy and sober-minded didacticism can be traced, Lanham argues, to the co-presence of contrary forms of selfhood, one philosophically serious, the other rhetorically playful. The courtier's ability to teach, delight, and move his audience depends on "keeping the two kinds of self in fruitful collision" so that he is always "moving from purpose to game . . . as circumstances require." As each new situation emerges, he must strike a necessarily provisional balance between complementary stances toward social reality. Didactic and speculative impulses endlessly revise and correct one another. The courtier's moral commitment must be weighed against his self-serving craving for personal authority, just as the productive spontaneity of immediate experience may issue merely in frivolous whimsy. Or to express the matter in a happy contradiction, one might say the ideal courtier will accede to the occasion of invention even as he seeks to change the world. True grace can therefore be defined as a process of continuous adjustment to and creation of the meaning of courtly life.[7]

7. Richard Lanham, *The Motives of Eloquence: Literary Rhetoric in the Renaissance* (New Haven: Yale University Press, 1976), pp. 32, 155. According to Lanham, Castiglione's rhetorical strategy in *The Book of the Courtier* facilitates our own acquisition of this ideal flexibility of response by continually unsettling reader preconceptions. If we "come to *The Courtier* with a serious referent reality . . . and a serious referent self, we shall find a narrative carefully prepared and consistently sustained as a game, a rhetorical contest." That is, if we expect Castiglione to pursue a moral objective, we will be surprised to discover his purpose is only aesthetic. Alternatively, if we "approach *The Courtier* with rhetorical coordinates . . . a curious transformation comes over it. The game of courtier, far from being one [the courtiers at Urbino] 'have never played,' becomes one that they play every day. The game becomes *realistic*, a straightforward mimesis of a dramatic reality." As readers,

Leonardo da Vinci is perhaps nowhere more gracefully inventive than in his sketches, which display all the spontaneity and improvisational verve Castiglione considered so crucial to the achievement of grace. The general effect of Leonardo's drawings, Clark says, "is one of a graceful speed and a hunter's certainty of eye."[8] As he takes aim at nature's multifariously moving targets, Leonardo's remarkably quick powers of observation enable him to endow even the most fleeting sensory impressions with pleasing aesthetic form. According to Freud, Leonardo's psychic apparatus was so responsibe to every stimulus he could not permanently commit himself to a single perspective on any object; instead, he constantly seeks fresh angles of insight. His sketches reveal, that is, "a wealth of possibilities between which a decision can only be reached with hesitation."[9] In a famous passage, Leonardo explicitly articulates this speculative, improvisational conception of invention when he advises his fellow painters to gaze at the stains on a wall, or "the embers of a fire, or clouds, or mud, or other similar objects from which you will find most admirable ideas . . . because from a confusion of shapes the spirit is quickened to new inventions."[10] Leonardo thus proposes an endless interpretive return to the occasion of thought; or as Ariel puts it in *The Tempest,* by "a sea-change into something rich and strange," he would render indifferent natural fact henceforth compatible with personal desire.

But Leonardo's interpretive mobility cannot, for all that, be equated with the playfully disinterested exuberance of an unfettered free spirit. His inventions are suffused rather with an air of restless striving, as if he were driven to invent by some inner compulsion. In contrast to the worldly eroticism of Titian's *Venus of Urbino,* which delights in the pleasures of the moment, the erotic for Leonardo is always impatient to find a new object, as in his painting of *Saint*

then, "we move continually in and out of seriousness, stay always in motion. We never adopt one relation to the world without lively awareness that we might have adopted the other." *The Book of the Courtier* thus inculcates a skill; precisely, the mobility in self-definition to which the assembled company at Urbino aspire; pp. 147, 148, 149.

8. Clark, *Leonardo da Vinci,* p. 31.
9. Freud, *Leonardo da Vinci, SE,* 11:68.
10. Quoted in Clark, *Leonardo da Vinci,* p. 82.

John, whose fluid turning motion and pointing finger gesture toward some undiscovered region beyond present appearances. Leonardo thus proposes that the truth, as it were, is nomadic, or forever on the move. Consider *The Adoration of the Kings,* a work whose unfinished character itself highlights Leonardo's desire not to be delayed by the routine elaboration of a project whose essential conception is already complete. The central figures in the composition, whose epochal and revelatory significance should subdue any audience in rapt wonderment, in fact appear in the midst of a swirling whirlpool of phantom shapes, as if only temporarily at the center of a vortex which might return them at any moment to a state of inchoate flux. According to Clark, "The symbolical homage of wisdom and science to a new faith is firmly expressed by the main figures; but pressing round them, like ghosts from the magical paganism of Apuleius, are those evasive creatures which writers on Leonardo are content to call angels. In the background, agitation of spirit inhabits the half-ruined construction of the intelligence." And on either margin, the two figures of "youth and age, moral and physical beauty, active and passive intelligence" that were for Leonardo powerful emblems of his own inventive will, at once responsive to experience and determined to impose form upon the world, these two figures stand apart, seeking their centers elsewhere.[11] Neither the mathematical decorum Florentine Neoplatonists discovered in antiquity, which promised coherent regularity in nature, a finite and therefore ultimately knowable universe, nor the potent familial symbolism of the Age of Faith with its redemptive vision of social harmony offers a representation of reality compelling enough to quell the anarchic dynamism of Leonardo's inventive will. His circumstantial intelligence could never finally assent to the proposition that humanity might transcend the vicissitudes of historical process by trusting to supposedly permanent truth. In his own words, "Whoever refers to authorities in disputing ideas, works with his memory rather than with his reason." Ever dissatisfied with the present shape of his beliefs, Leonardo relentlessly quests after unprecedented experience in order to forge an original albeit transitory synthesis.

11. Ibid., p. 40.

The restless, agitated nature of Leonardo's remarkable inventiveness suggests his determination to investigate reality is but the corollary of an equally acute vulnerability to the world's disruptive impact. Leonardo was in fact perpetually traumatized by his surroundings. The graceful continuum of flowing water he may have intuitively recognized as a compelling correlative for his own continuously inventive intelligence at every instant verges perilously close to an uncontrollable flood. His ego was not a barricade behind which he could retreat when confronted with novel experience. Instead, it might better be thought of as a permeable membrane so open to penetration from without that Leonardo was always flooded by desires demanding satisfaction. Kurt Eissler observes in this connection that "Leonardo's stream of graphic productions" issues from his fear of death, which "amounts to a feeling that the self is under a constant threat of disorganization." His everyday interaction with his environment ceaselessly provoked in him an "accumulation of longings which meant for Leonardo an acceleration towards death," an acceleration, that is, toward the dissolution of the self. From moment to moment thrown into a state of panic, Leonardo sought desperately to reestablish his precarious sense of self-possession by subduing the world symbolically, by converting his present experience into an idea or pictorial representation. As Eissler says, "In the genius the loss of structure leads to a hypertrophy of function."[12] Leonardo's countless inventions were inspired by his urgent need to appease the desires proliferating within him and so remove the threat of psychic disorganization. His gracefully fluid improvisational skill was employed in the service of an interminable effort to achieve a condition of invulnerable mastery. In a striking indeed exorbitant manner, Leonardo's career reveals the underlying motivation of graceful comportment.

Freud's famous analysis of Leonardo's infantile reminiscence is pertinent here. Freud interprets Leonardo's recollection of a dream in which a great bird descends upon him, striking repeatedly with its tail against his lips, as proof of an especially intense maternal relationship: "The phantasy is compounded from the memories of being

12. Eissler, *Leonardo da Vinci: Psychoanalytic Notes on the Enigma* (New York: International Universities Press, 1961), pp. 219, 224, 218.

suckled and of being kissed by the mother."[13] That is, when Leonardo was a child, the love he felt for his mother was strongly reinforced by the doting attention she lavished upon him. This profound attachment to his maternal love object was decisive for Leonardo's subsequent development. By a process of sublimation, it was generalized to encompass all of reality, so that the child's desiring interest in his mother became the basis for the adult's eventually scientific interest in (mother) nature. But what is crucial about Leonardo's fantasy is its passive, masochistic character. In this recollection of what can only be described as the infant's primordial seduction, the birth of desire transpires as aggression directed toward the subject: the child has been invaded by the world. A disturbing sensation of ontological insecurity thus replaces the psychic equilibrium of prenatal existence. The ominous undertone Walter Pater detected in the *Mona Lisa*'s smile can perhaps be explained in this connection: to the extent that the *Mona Lisa* embodies Leonardo's memory of his own mother, one might say his mother's smile, sign of her affectionate regard, simultaneously conveyed to him a "sinister menace," as Freud puts it, because her "promise of unbounded tenderness" created a condition of disorienting psychically painful excitement in the mind of an infant as yet unprepared to defend itself against intrusions from without.[14]

In retaliation, as it were, for this unsettling introduction to sexuality, the young Leonardo then went on to conceive the project of subduing his surroundings so as to remove the possibility of further traumatic encounters. He thus produced an oral sadistic fantasy of consuming the totality of existence not simply in order to obtain an increased yield in pleasure but to ward off the terrifying possibility of personal dissolution. According to Freud, Leonardo's infantile fantasy of having the world within himself later served as an important motive force propelling the intellectual activity of adulthood. That is, the adult's famous curiosity about the nature of reality is really a sublimation of the child's cannibalistic rage. Leonardo's fall from grace, his obsessive need to know everything can be compared in this sense to the rapacity and lust for power animating the mega-

13. Freud, *SE*, 11:107.
14. Ibid., p. 115.

lomaniac Kurtz in Joseph Conrad's *Heart of Darkness.* Conrad's narrator Marlow describes Kurtz precisely in oral terms: Kurtz "opened his mouth wide—it gave him a weirdly voracious aspect, as though he had wanted to swallow all the air, all the earth, all the men before him." Kurtz's primitive desire literally to devour reality is, of course, the motive for his eloquence—for his "gift of expression, the bewildering, the illuminating, the most exalted and the most contemptible, the pulsating stream of light, or the deceitful flow from the heart of an impenetrable darkness"—just as Leonardo's inventiveness issues from his own wish to ingest the world.[15] A second meaning of the Leonardesque smile thus emerges: to the extent that the *Mona Lisa* represents Leonardo's ideal self, one might say her smile signifies the imperturbable self-satisfaction of a being who has succeeded in completely consuming every possible object of desire.

The fantasy of consummate satisfaction figures prominently in Leonardo's imaginative life throughout his career. For instance, it influences his conception of science, which he defines as the systematic accumulation of knowledge proceeding "in orderly sequence from the first true and established principles through successive stages to the end," at which point the investigator will have achieved a comprehensive picture of the entire universe. Leonardo is therefore hostile to a rhetorical perspective emphasizing the productive role disputation plays in the development of scientific thought: "Where the cry of controversy is heard, there is no true science, because the truth has one single end and when this end is published, argument is destroyed forever." He hoped the mathematization of nature would finally silence the "tongues of argument," "the lying sciences of the mind," those "sophistical sciences which lead to interminable conflict." Leonardo esteems the art of painting for the same reason he advises us to rely on the "supreme certainty of mathematics": like mathematics, painting also holds forth the possibility of permanent, total knowledge. Clark notes that Leonardo preferred painting to music because "its sequences are not fleeting sounds or images . . . but can be apprehended immediately and contemplated

15. Joseph Conrad, *Heart of Darkness and The Secret Sharer* (New York: Doubleday, 1950), pp. 135, 119–20.

indefinitely."[16] Or as Freud suggests, when Leonardo began to paint, he was unable "to limit his demands, to see the work of art in isolation and to tear it from the wide context to which he knew it belonged." Rather, he sought "to bring to expression in it everything that was connected with it in his thoughts."[17] This desire for totalization culminates in the Deluge drawings. In them, Leonardo encounters the wellspring of his creativity in the form of a cataclysmic flood of desiring impulses inundating the earth. Instead of gradually coming to know the world, the Deluge drawings enact a fantasy of accelerating metamorphosis in which the desiring subject knows, indeed, *becomes* everything at once. These drawings are Leonardo's apocalypse. They present a spectacle which, for Leonardo, was simultaneously terrifying and exhilarating: the sublime prospect of the end of desire.

Yet even as he yields to a fantasy of totalization, Leonardo begins to move away from it. According to Clark, Leonardo is obsessed with his premonition of the end of the world at the beginning of the Deluge series, but in the later drawings he becomes "so absorbed in the elaboration of curvilinear patterns that the scenes lose some of their dramatic force."[18] As the series unfolds, Leonardo's vision of desiring apocalypse is gradually displaced from the center of his imagination, which now turns to the task of pictorial representation. Thus, in the sketch selected as frontispiece for this essay, Leonardo explores the purely formal, aesthetic possibilities of the Deluge theme. The Deluge, Clark says, seems here "to have lost some of its ferocity of purpose, the clouds and waterspouts have almost the harmless decorative effect of conventional waves on a Korin screen."[19] Leonardo refuses, that is, to take the Deluge seriously: he retreats from the prospect of perfect mastery, steps back from the abyss of suicidal desire. Even as the ethic of shapeshifting constitutes a privileged point of reference, Leonardo resists becom-

16. Quoted in Clark, *Leonardo da Vinci,* pp. 84–85, 64; see p. 75.
17. Freud, *SE,* 11:77.
18. Clark, *Leonardo da Vinci,* p. 153.
19. Kenneth Clark, *The Drawings of Leonardo da Vinci,* 2d ed. (London: Phaidon, 1968–69), p. 55.

ing trapped in a narcissistic self-regard. The spectacle of climactic satisfaction fails to quench the eternal flame of his inventive will. Instead, desire demonstrates its infinitely renewable character as Leonardo recurs from his fantasy of unlimited bliss to the provisional pleasures of the moment. In this reaffirmation of phenomenal reality, Leonardo converts his fantasized identity into a project, an activity of invention wandering forever in that indeterminate middle zone where interpretation is truly endless. Leonardo thus consents to a partial, contingent form of selfhood that never achieves a conclusive stage of realization.

On the one hand, then, there is the desire to possess the meaning of the world once and for all; on the other, its subversion. Leonardo's wish to be done with invention has itself become the occasion for improvisation and experiment. To put it another way, if the will-to-invent cannot finally be separated from the desire for finality, Leonardo's Deluge series, representing the apotheosis of his inventive will, nevertheless becomes mobile with respect to itself through a foregrounding of pictorial means. This pictorial irony is a saving grace. It signifies Leonardo's ironic stance toward his conception of himself as *magister ludi.* Leonardo thus withdraws assent from the attempt to transform historical process into a static tableau expressing the essence of the will-to-invent's endless productivity. His fantasy that perfect satiation might be achieved through a protean diversification of desire openly proclaims its fantasmatic status. The saving grace resides precisely in the fact that this is only an *image* of change; an image of change providing an analogy for the remarks with which the present essay will now conclude.

Kaironomia has been an encomium, or formal praise, of the will-to-invent in four parts corresponding to the traditional itinerary of Introduction, Argument, Proof, and Conclusion. It has sought both to characterize and to demonstrate an ideally circumstantial or improvisational activity of invention. In the introductory section, the concept of *kairos* was invoked as precedent for the Argument subsequently made on behalf of contemporary aspirations for a speculatively mobile form of interpretation. The ensuing Proof then provided a practical demonstration of interpretive mobility. *Kaironomia*'s Introduction, Argument, and Proof thus enact the gradual achievement of an ideal, moving from past to present and from theory to practice.

But the simultaneous praise and blame of Leonardo's Deluge drawings that has taken up the greater portion of *Kaironomia*'s Conclusion implicitly serves notice of the fraudulent nature of this ascent toward perfection. That is, the proof section presents a dissembling of occasionality, an aesthetically shaped fiction of hermeneutical process. It makes the prospect of change pleasurable by contriving a fair semblance of speculative thought. The demonstration of interpretive mobility transpires, in other words, within a closed field of possibilities, a field of possibilities defined by the preceding sections of *Kaironomia.* Part 3 essentially reiterates Parts 1 and 2. Thus, "the eternally renascent phallus, sign of perpetual desire" is an equivalent expression for "mutability as the leading characteristic of reality." "The unending strife of fraternal contraries" corresponds to "the antithetical nature of truth." And "the self-satisfied complacency of patriarchal narcissism" can properly be associated with "the deluded belief that persuasion rests on the representational efficacy of language." In fact, throughout the course of its exposition, *Kaironomia* has merely amplified its own beginning premises. The recurrent pairs of opposing terms—"beauty and sublimity"; "dogmatism and doubt"; "tragedy and farce"—all rhyme, as it were, with the fundamental contradiction embodied in *Kaironomia*'s title. Similarly, "inventing the meaning of the world on the spur of the moment," "the middle voice as playfully endless interpretation," "sublimation on the basis of unprecedented desire"—all these formulations are synonyms for a single way of being in the world, for the stance toward experience, the *ethos* of "the sophist," "the shaman," "the ideal courtier."

To put the matter briefly, *Kaironomia*'s programmatic recourse to select literary and philosophic works of antiquity, the Renaissance, the Romantic and modern periods, its catalogue of exemplary cultural figures—Gorgias and Lao-tze; Leonardo and Castiglione; Coleridge and Blake; Freud, Nietzsche, Joyce, and Stein; Barthes, Foucault, and so on—bespeaks a covert aspiration for timeless authority. From the beginning, the desire to promote a practice of speculative thought alert to its own occasion has been undermined by a contrary impulse which would transcend history in order to achieve a conclusive, enduring form. The inaugural opposition of *kairos* and *nomos,* spontaneity and tradition, was already resolved in favor of the latter from the moment *kairos* was defined as a radical principle of occasionality establishing the living present as point of departure for rhetorical invention. This essay's trajectory could have been predicted at the outset. Instead of progressive disclosure of a hitherto concealed truth, *Kaironomia* has unfolded rather as a suite of variations on a theme. It is itself an instance of *kaironomia,* a provisional and revocable attempt to make sense; precisely, a historically contingent management of the opportunity to write an encomium on the will-to-invent. But then, the will-to-invent can renew itself only by withdrawing assent from its latest production. As a final gesture, *Kaironomia* thus seeks to perform a vanishing act, so that the effect of these concluding remarks would be like that of a slipknot.

Works Cited

Adkins, A. W. *Merit and Responsibility: A Study in Greek Values.* Oxford: Clarendon Press, 1960.

Allen, Donald, ed. *The Poetics of the New American Poetry.* New York: Grove Press, 1973.

Aristotle. *The Works of Aristotle.* Translated into English under the editorship of W. D. Ross. Oxford: Clarendon Press, 1908–62.

Barber, C. L. *Shakespeare's Festive Comedy.* Princeton: Princeton University Press, 1959.

Barthes, Roland. *Critical Essays.* Translated by Richard Howard. Evanston, Ill.: Northwestern University Press, 1972.

______. *Image-Music-Text.* Essays selected and translated by Stephen Heath. New York: Hill and Wang, 1977.

______. *The Pleasure of the Text.* Translated by Richard Miller. New York: Hill and Wang, 1975.

______. *S/Z.* Translated by Richard Miller. New York: Hill and Wang, 1974.

Beckett, Samuel, et al. *Our Exagmination Round His Factification for Incamination of Work in Progress.* 1929; rpt. New York: New Directions, 1962.

Bellah, Robert. *Beyond Belief: Essays on Religion in a Post-Traditional World.* New York: Harper and Row, 1970.

Benveniste, Emile. *Problems in General Linguistics.* Translated by M. E. Meek. Coral Gables, Fla.: University of Miami Press, 1971.

Bersani, Leo. *Baudelaire and Freud.* Berkeley and Los Angeles: University of California Press, 1977.

______. *A Future for Astyanax: Character and Desire in Literature.* Boston: Little, Brown, 1976.

BLAKE, WILLIAM. *Complete Poetry and Prose*. Edited by Geoffrey Keynes. London: Nonesuch Library, 1975.

BRIDGMAN, RICHARD. *Gertrude Stein in Pieces*. New York: Oxford University Press, 1970.

BROWN, NORMAN O. *Closing Time*. New York: Random House, 1973.

———. *Life against Death: The Psychoanalytical Meaning of History*. Middletown, Conn.: Wesleyan University Press, 1959.

BRUNS, GERALD L. *Inventions: Writing, Textuality, and Understanding in Literary History*. New Haven: Yale University Press, 1982.

CABANNE, PIERRE. *Dialogues with Marcel Duchamp*. Translated by Ron Padgett. New York: Viking Press, 1971.

CAMPBELL, JOSEPH, AND HENRY MORTON ROBINSON. *A Skeleton Key to Finnegans Wake*. New York: Harcourt, Brace, 1944.

CASTANEDA, CARLOS. *A Separate Reality*. Harmondsworth, Middlesex: Penguin, 1973.

CASTIGLIONE. *The Book of the Courtier*. Translated by Charles S. Singleton. New York: Anchor, 1959.

CAWS, MARY ANN. *The Poetry of Dada and Surrealism*. Princeton: Princeton University Press, 1970.

CHABROL, CLAUDE, ED. *Sémiotique narrative et textuelle*. Paris: Larousse, 1973.

CHAN, WING-TSIT. *A Source Book in Chinese Philosophy*. Princeton: Princeton University Press, 1963.

CHANG CHUNG-YUAN. *Tao: A New Way of Thinking: A Translation of the Tao Te Ching with an Introduction and Commentaries*. New York: Harper and Row, 1977.

CHATMAN, SEYMOUR. "The Rhetoric of Difficult Fiction." *Poetics Today* 1, no. 4 (Summer 1980): 23–66.

CHENG, CHUNG-YING. "Nature and Function of Scepticism in Chinese Philosophy." *Philosophy East and West* 27, no. 2 (April 1977): 137–54.

CIXOUS, HELENE. *The Exile of James Joyce*. Translated by Sally A. J. Purcell. New York: D. Lewis, 1972.

CLARK, KENNETH. *The Drawings of Leonardo da Vinci in the Collection of Her Majesty the Queen at Windsor Castle*. 2d ed., rev. with the assistance of Carlo Pedretti. London: Phaidon, 1968–69.

———. *Leonardo da Vinci: An Account of His Development as an Artist*. Harmondsworth, Middlesex: Penguin, 1967.

COLERIDGE, SAMUEL TAYLOR. *Biographia Literaria*. Edited with an introduction by George Watson. London: J. M. Dent and Sons, 1971.

———. *Coleridge on Shakespeare*. Edited by Terence Hawkes with an introduction by Alfred Harbage. Harmondsworth, Middlesex: Penguin, 1969.

———. *Treatise on Method as Published in the Encyclopedia Metropolitana*. Edited by Alice Snyder. London: Constable, 1934.

CONRAD, JOSEPH. *Heart of Darkness and The Secret Sharer.* With an introduction by Albert J. Guerard. New York: Doubleday, 1950.

CORTAZAR, JULIO. *Hopscotch.* Translated by Gregory Rabassa. New York: Random House, 1966.

CULLER, JONATHAN. *Structuralist Poetics: Structuralism, Linguistics, and the Study of Literature.* Ithaca: Cornell University Press, 1975.

CURTIUS, E. R. *European Literature and the Latin Middle Ages.* Translated by W. R. Trask. New York: Pantheon, 1953.

DARWIN, CHARLES. *The Origin of Species by Means of Natural Selection.* Edited with an introduction by J. W. Burrow. Harmondsworth, Middlesex: Penguin, 1968.

DAWSON, RAYMOND, ED. *The Legacy of China.* Oxford: Oxford University Press, 1964.

DE MAN, PAUL. *Blindness and Insight: Essays in the Rhetoric of Contemporary Criticism.* New York: Oxford University Press, 1971.

_____. "Nietzsche's Theory of Rhetoric." *Symposium* 28, no. 1 (1974): 33–51.

DE ROMILLY, JACQUELINE. *Magic and Rhetoric in Ancient Greece.* Cambridge, Mass.: Harvard University Press, 1975.

DERRIDA, JACQUES. *Of Grammatology.* Translated by Gayatri Chakravorty Spivak. Baltimore: Johns Hopkins University Press, 1976.

_____. *Speech and Phenomena.* Translated by David Allison. Evanston, Ill.: Northwestern University Press, 1973.

DETIENNE, MARCEL, AND JEAN-PIERRE VERNANT. *Cunning Intelligence in Greek Culture and Society.* Translated by Janet Lloyd. Atlantic Highlands, N.J.: Humanities Press, 1978.

DONOGHUE, DENIS. *Ferocious Alphabets.* Boston: Little, Brown, 1981.

DONOGHUE, DENIS, AND J. R. MULRYNE, EDS. *An Honoured Guest: New Essays on W. B. Yeats.* London: Edward Arnold, 1965.

DREYFUS, HUBERT L., AND PAUL RABINOW. *Michel Foucault: Beyond Structuralism and Hermeneutics.* 2d ed. With an Afterword by and an interview with Michel Foucault. Chicago: University of Chicago Press, 1983.

EISSLER, KURT. *Leonardo da Vinci: Psychoanalytic Notes on the Enigma.* New York: International Universities Press, 1961.

ELIADE, MIRCEA. *Shamanism: Archaic Techniques of Ecstasy.* Translated by W. R. Trask. New York: Pantheon, 1964.

ELLMANN, RICHARD. *James Joyce.* New York: Oxford University Press, 1959.

FOUCAULT, MICHEL. *The History of Sexuality.* Vol. I: *An Introduction.* Translated by Robert Hurley. New York: Random House, 1978.

_____. *The Order of Things: An Archaeology of the Human Sciences.* London: Tavistock, 1970.

FRAZER, JAMES. *The New Golden Bough: A New Abridgment of the Clas-*

sic Work. Edited, and with notes and a foreword, by Theodor H. Gaster. New York: Criterion, 1959.

FREUD, SIGMUND. *The Standard Edition of the Complete Psychological Works of Sigmund Freud*. Under the general editorship of James Strachey. 24 vol. London: Hogarth Press, 1953–74.

GADAMER, HANS-GEORG. *Truth and Method*. Translated by G. Banden and J. Cumming. New York: Seabury Press, 1975.

GLASHEEN, ADALINE. *A Third Census of Finnegans Wake*. Berkeley and Los Angeles: University of California Press, 1977.

GLASS, PHILIP. Liner Notes. *Music in Twelve Parts, Parts 1 & 2*. Virgin Records, CA2010, 1974.

GOODY, JACK, AND IAN WATT. "The Consequences of Literacy." *Comparative Studies in Society and History* 5 (1963): 305–45.

GOULD, STEPHEN JAY. *The Mismeasure of Man*. New York: Norton, 1981.

GRUBE, G. M. A. *The Greek and Roman Critics*. London: Methuen, 1965.

GUTHRIE, W. K. C. *The Sophists*. Cambridge: Cambridge University Press, 1971.

HARARI, JOSUÉ V., ED. *Textual Strategies: Perspectives in Post-Structuralist Criticism*. Ithaca: Cornell University Press, 1979.

HARTMAN, GEOFFREY. *Criticism in the Wilderness: The Study of Literature Today*. New Haven: Yale University Press, 1980.

______. *Saving the Text: Literature/Derrida/Philosophy*. Baltimore: Johns Hopkins University Press, 1981.

HEATH, STEPHEN. "Ambiviolences: Notes pour la lecture de Joyce." *Tel Quel* 50 & 51 (1972): 22–43, 64–76.

______. *The Nouveau Roman: A Study in the Practice of Writing*. London: Elek, 1972.

HOLMBERG, UNO. *Finno-Ugric, Siberian Mythology*. Vol. 4 of *The Mythology of All Races*. Edited by Louis Herbert Gray. New York: Cooper Square Publishers, 1964.

HUSSERL, EDMUND. *Cartesian Meditations: An Introduction to Phenomenology*. Translated by Dorion Cairns. The Hague: Martinus Nijhoff, 1977.

______. *The Crisis of European Sciences and Transcendental Phenomenology: An Introduction to Phenomenological Philosophy*. Translated, with an introduction, by David Carr. Evanston, Ill.: Northwestern University Press, 1970.

JAMES, WILLIAM. *The Principles of Psychology*. Vol. 1. 1890; rpt. New York: Dover, 1950.

JAVITCH, DANIEL. *Poetry and Courtliness in Renaissance England*. Princeton: Princeton University Press, 1978.

JOHNSON, BARBARA. *The Critical Difference: Essays in the Contemporary Rhetoric of Reading*. Baltimore: Johns Hopkins University Press, 1980.

JONES, JOHN. *On Aristotle and Greek Tragedy.* London: Chatto and Windus, 1962.
JOYCE, JAMES. *Finnegans Wake.* 3d ed. London: Faber and Faber, 1975. (First edition, 1939; 3d ed. 1964.)
KENNEDY, GEORGE. *The Art of Persuasion in Greece.* Princeton: Princeton University Press, 1963.
KERFERD, G. B. "Aristotle." In *The Encyclopedia of Philosophy.* Edited by Paul Edwards. New York: Macmillan, 1967. Vol. 1: 151–62.
KERMODE, FRANK. *The Genesis of Secrecy: On the Interpretation of Narrative.* Cambridge, Mass.: Harvard University Press, 1979.
KLEIN, MELANIE. *Contributions to Psychoanalysis, 1921–1945.* Introduction by Ernest Jones. London: Hogarth Press, 1948.
KOCKELMANS, JOSEPH J., ED. *Phenomenology: The Philosophy of Edmund Husserl and Its Interpretation.* Garden City, N.Y.: Anchor, 1967.
LACAN, JACQUES. *Ecrits.* Paris: Editions de Seuil, 1966.
______. *Ecrits: A Selection.* Translated by Alan Sheridan. New York: Norton, 1977.
LANHAM, RICHARD. *The Motives of Eloquence: Literary Rhetoric in the Renaissance.* New Haven: Yale University Press, 1976.
LAO-TZE. *Tao Te Ching.* Translated by Gia-fu Feng and Jane English. London: Wildwood House, 1973.
LAPLANCHE, JEAN. *Life and Death in Psychoanalysis.* Translated by Jeffrey Mehlman. Baltimore: Johns Hopkins University Press, 1976.
______. *Problématiques III: La sublimation.* Paris: Presses Universitaires de France, 1980.
LEACH, EDMUND. *"Genesis as Myth" and other Essays.* London: Jonathan Cape, 1969.
LYOTARD, JEAN-FRANÇOIS. *The Postmodern Condition: A Report on Knowledge.* Translated by Geoff Bennington and Brian Massumi with a foreword by Fredric Jameson. Minneapolis: University of Minnesota Press, 1984.
MACCABE, COLIN. *James Joyce and the Revolution of the Word.* London: Macmillan, 1978.
MCFARLAND, THOMAS. *Originality and Imagination.* Baltimore: Johns Hopkins University Press, 1985.
MCHUGH, ROLAND. *Annotations to Finnegans Wake.* Baltimore: Johns Hopkins University Press, 1980.
MACKSEY, RICHARD, AND EUGENIO DONATO, EDS. *The Structuralist Controversy: The Languages of Criticism and the Sciences of Man.* Baltimore: Johns Hopkins University Press, 1970.
MAGOUN, FRANCIS PEABODY, JR., TRANS. *The Kalevala or Poems of the Kalevala District.* Compiled by Elias Lönnrot. Cambridge, Mass.: Harvard University Press, 1963.

MEHLMAN, JEFFREY. *A Structural Study of Autobiography: Proust, Leiris, Sartre, Lévi-Strauss.* Ithaca: Cornell University Press, 1974.

MONTAIGNE. *Essays and Selected Writings.* A bilingual edition translated and edited by Donald M. Frame. New York: St. Martin's Press, 1963.

MOTHERWELL, ROBERT, ED. *The Dada Painters and Poets.* 1951; rpt. New York: G. K. Hall, 1981.

NIETZSCHE, FRIEDRICH. *The Birth of Tragedy and the Genealogy of Morals.* Translated by F. Golffing. Garden City, N.Y.: Doubleday, 1956.

______. *Early Greek Philosophy and Other Essays.* Translated by M. A. Mügge. Vol. 2 of *The Complete Works of Friedrich Nietzsche.* Edited by Dr. Oscar Levy. New York: Macmillan, 1909–11.

______. *The Gay Science: With a Prelude in Rhymes and an Appendix of Songs.* Translated, with commentary, by Walter Kaufmann. New York: Random House, 1974.

______. *The Portable Nietzsche.* Translated and edited by Walter Kaufmann. Revised edition. London: Chatto and Windus, 1971.

______. *The Will to Power.* A new translation by Walter Kaufmann and R. J. Hollingdale. Edited, with commentary, by Walter Kaufmann. New York: Random House, 1967.

NORRIS, MARGOT. *The Decentered Universe of Finnegans Wake: A Structuralist Analysis.* Baltimore: Johns Hopkins University Press, 1976.

OINAS, FELIX J. "The Finnish and Estonian Folk Epic." *Journal of Baltic Studies* 7, no. 1 (Spring 1976): 1–12.

ONIANS, R. B. *The Origins of European Thought about the Body, the Mind, the Soul, the World, Time, and Fate: New Interpretations of Greek, Roman and Kindred Evidence, Also of Some Basic Jewish and Christian Beliefs.* Cambridge: Cambridge University Press, 1951.

OVID. *Metamorphoses.* Translated by Rolfe Humphries. Bloomington: Indiana University Press, 1955.

PERNIOLA, MARIO. "Logique de la séduction." Translated by Thierry O. Séchan. *Traverses* 18 (February 1980): 2–9.

POIRIER, RICHARD. *The Performing Self: Compositions and Decompositions in the Languages of Contemporary Life.* New York: Oxford University Press, 1971.

POPPER, KARL. *The Open Society and Its Enemies.* Princeton: Princeton University Press, 1950.

POUND, EZRA, *The Cantos of Ezra Pound.* New York: New Directions, 1972.

POWELL, T. G. E. *Prehistoric Art.* New York: Praeger, 1966.

RANTA, TAIMA. "Finnish Folklore." *Michigan Academician* 3, no. 3 (Winter 1971): 23–40.

RICOEUR, PAUL. *The Rule of Metaphor: Multi-disciplinary Studies in the Creation of Meaning in Language.* Translated by Robert Czerny with

Kathleen McLaughlin and John Costello. Toronto: University of Toronto Press, 1977.

RONAN, COLIN A. *The Shorter Science and Civilization in China: An Abridgement of Joseph Needham's Original Text.* Vol. 1. Cambridge: Cambridge University Press, 1978.

RORTY, RICHARD. *Philosophy and the Mirror of Nature.* Princeton: Princeton University Press, 1979.

SAID, EDWARD. *Beginnings: Intention and Method.* New York: Basic Books, 1975.

SAUSSURE, FERDINAND DE. *Course in General Linguistics.* Edited by Charles Bally and Albert Sechehaye; with the collaboration of Albert Riedlinger. Translated and annotated by Roy Harris. London: Duckworth, 1983.

SCHILLER, FRIEDRICH. *Letters on the Aesthetic Education of Man.* Translated by Reg Snell. New Haven: Yale University Press, 1954.

SOLLERS, PHILIPPE. "Joyce & Co." Translated by Stephen Heath. *Triquarterly* 38 (1977): 107–21.

SOLOMON, MARGARET. *Eternal Geomater: The Sexual Universe of Finnegans Wake.* Carbondale: Southern Illinois University Press, 1969.

SPRAGUE, ROSAMOND KENT, ED. *The Older Sophists.* Columbia: University of South Carolina Press, 1972. A translation of Diels-Kranz, eds., *Die Fragmente der Vorsokratiker* (1934).

STEIN, GERTRUDE. *Everybody's Autobiography.* New York: Random House, 1937.

______. *Lectures in America.* New York: Random House, 1935.

______. *The Selected Writings of Gertrude Stein.* Edited, with an introduction and notes, by Carl Van Vechten; with an essay on Gertrude Stein by F. W. Dupee. New York: Random House, 1962.

STRAVINSKY, IGOR. *Chamber Works.* Cond. Pierre Boulez, Ensemble Intercontemporain Deutsche Grammaphon, 2531–378, 198[illegible]

UNTERSTEINER, MARIO. *The Sophists.* Translated by K. Freeman. Oxford: B. Blackwell, 1954.

VICKERY, JOHN B. "*Finnegans Wake* and Sexual Metamorphosis." *Contemporary Literature* 13, no. 2 (Spring 1972): 213–42.

VICO, GIAMBATTISTA. *New Science.* Revised translation of the 3d edition (1774) by Thomas Goddard Bergin and Max Harold Frisch. Ithaca, N.Y.: Cornell University Press, 1968.

WILSON, WILLIAM A. "The *Kalevala* and Finnish Politics." *Indiana University Folklore Institute Journal* 12 (1975): 131–55.

Index

Library of Congress Cataloging-in-Publication Data

White, Eric Charles, 1952–
Kaironomia : on the will-to-invent

Bibliography: p.
Includes index.
1. Invention (Rhetoric) 2. Persuasion (Rhetoric)
I. Title.
PN221.W48 1987 808 86-24235
ISBN 0-8014-1993-X (alk. paper)